DESIGN FOR GOOD ACOUSTICS AND NOISE CONTROL

Design for Good Acoustics and Noise Control

J. E. Moore, F.R.I.B.A.

First published 1978 by
THE MACMILLAN PRESS LTD
London and Basingstoke
Associated companies in Delhi Dublin
Hong Kong Johannesburg Lagos Melbourne
New York Singapore and Tokyo

Typeset in 10/11 pt IBM Press Roman by
Reproduction Drawings Ltd, Sutton, Surrey
and printed in Hong Kong by
Brighter Printing Press Ltd.

British Library Cataloguing in Publication Data

Moore, John Edwin
Design for good acoustics and noise control.
1. Architectural acoustics
I. Title
729'.29 NA2800

ISBN 0-333-24292-0
ISBN 0-333-24293-9 Pbk

Contents

Preface

This book has been written primarily as a textbook for students of architecture, interior design, town planning and surveying. With this broad purpose in mind I have attempted to provide in one volume of reasonable size as comprehensive a survey as possible of those acoustic factors which affect the design of rooms, buildings and urban development. This has meant dealing somewhat briefly with some aspects of the subject while other areas are discussed in more depth. The fairly extensive bibliography should, however, direct the student to further sources of study in those cases where space did not allow of a more detailed treatment. For example, I considered that an introduction to noise level contouring was all that was necessary since this technique is very fully described in inexpensive publications issued by the Department of the Environment.

I have also tried to arrange the material in a way that will make subsequent reference easy. There is first of all the broad division into four sections

(1) Properties and behaviour of sound
(2) Subjective aspects of sound
(3) Noise control
(4) Room acoustics.

In the last two sections principles and calculations are separated, again for easy reference. Calculations have been limited to those which I think the student will find most useful in design and which typify the methods employed.

Within this broad framework each subsidiary item has a clear subheading in the text and these are listed on a contents page at the beginning of each section. These pages, taken together, form an index that may often be more useful than the orthodox index at the end of the book. For example, it is possible for the reader to refresh his memory of a given item by following the thread of the subject through the four contents pages. If we take the phenomenon of 'screening' as a case in point, the reader will find this developed under the following subheadings:

Diffraction and sound shadows. Reduction of internal noise by screening. Reduction of external noise by screening. Screening by planting. Effect of screening, single-figure calculations. Effect of screening, six-figure calculations. Acoustics for speech, sound shadows.

Finally, I have not assumed any previous knowledge of the physics of sound, building acoustics or even logarithmic scales. Many students will therefore find the earlier part of the book somewhat elementary but I considered it best not to allow any possible gaps in knowledge to hinder the understanding of the rest of the book.

J. E. MOORE

Acknowledgements

The author wishes to thank P. H. Parkin for permission to reproduce the graph in figure 3.40. He is also very much indebted to the authors of many of the publications listed in the bibliography which were consulted during the preparation of this book. The schedules in table 3.14 are reproduced by permission of the Director of Her Majesty's Stationery Office.

1 Properties and Behaviour of Sound

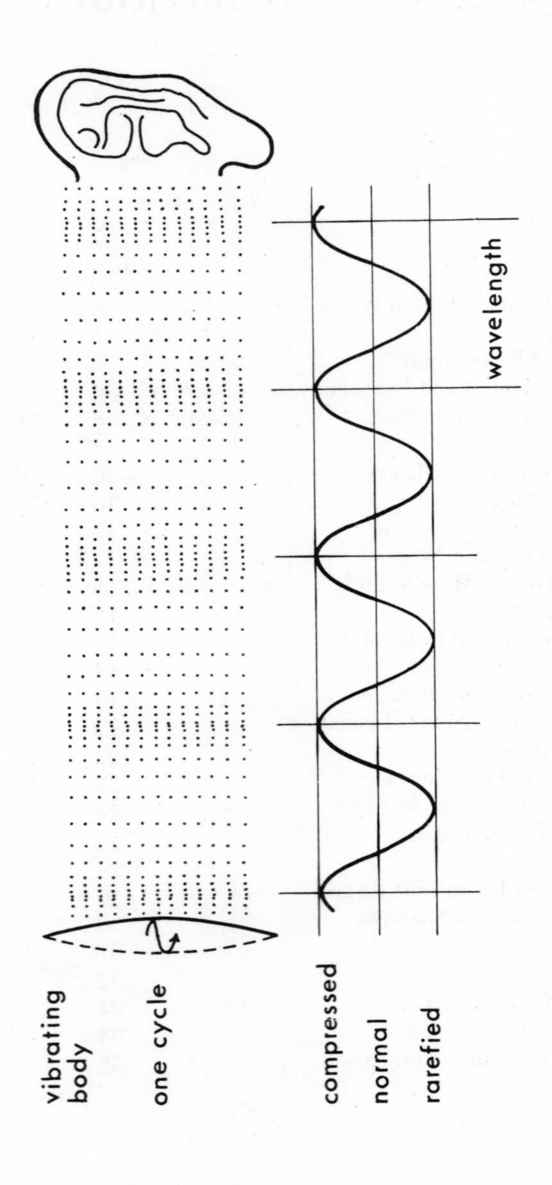
vibrating body
one cycle
compressed
normal
rarefied
wavelength

Figure 1.1

Definition of Sound

Sound may be defined as vibrations or pressure changes in an 'elastic' medium which are capable of being detected by the ear. By 'elastic' we mean that the particles of the medium return to their original position after disturbance by the vibrational wave. Such vibrations travel through solids, liquids and gases but the normal process of hearing depends on their ultimate transmission through the air so that the ear drum is set in vibration and a sequence of events we call 'hearing' begins.

The above definition can, however, be widened to include sonic vibrations which, because of their very high or very low frequency, cannot be detected by the human ear. In architectural acoustics we will not be concerned with vibrations of this nature.

In the subject of room acoustics we are mainly concerned with sound vibrations transmitted through the air, whereas in the subject of noise control we are equally concerned with vibrations transmitted by way of solid materials. The path of sound is frequently through both media before it finally reaches the listener.

Source of Sound

The origin of sound vibrations is usually the vibration of a solid body caused by the application of physical energy, such as when a harp string is plucked or when the rotary motion of a machine causes vibrations in its associated parts. Sound waves can also be caused by air turbulence, by an explosive expansion of air, or by a combination of such events. If, however, we consider the simple case of a vibrating solid body, the general characteristics of sound waves can be more readily understood.

Sound Waves in Air

Figure 1.1 represents diagrammatically a vibrating element such as a loudspeaker cone. It is emitting a 'pure tone', that is, a simple vibration of one frequency. The outward movement of the diaphragm causes a compression in the adjoining particles of air. When the diaphragm moves back the compressed air expands and causes a compression in the air adjoining the original centre of compression. This compression in turn disturbs the air in advance of it, and so on. What is called a 'travelling wave' is set up. Thus the original compression travels outwards, followed by further compressions as the diaphragm continues to vibrate. The drawing shows a series of such compressions arrested at a point in time. When these fluctuating pressure waves impinge on the ear drum it, in turn, vibrates and the process of hearing begins. The air between source and listener does not move bodily. At any point along the sound path the particles of air are merely moving from side to side of their original position. It is only the disturbance, or wave, which travels. The air may in certain situations be moving bodily, such as in a ventilation duct; nevertheless, the sound wave will move outwards from the source regardless of the direction of air movement.

The graph below the drawing in figure 1.1 shows the changes in pressure along the path between source and listener. The central horizontal line represents air at normal atmospheric pressure. The wave form represents the increase and decrease of pressure due to the sound wave, that is, the amount plus or minus atmospheric pressure. For sounds of average loudness this deviation is only about one millionth of normal atmospheric pressure. Thus if the rise and fall of the curve of the graph measured 1 cm, the distance to the base of the graph (zero pressure) would be 10 km or about 6 miles.

The General Form of Sound Waves

A vibrating surface, such as the one considered above, will emit sound waves not only in front and behind the surface but also, by diffraction, in all directions.

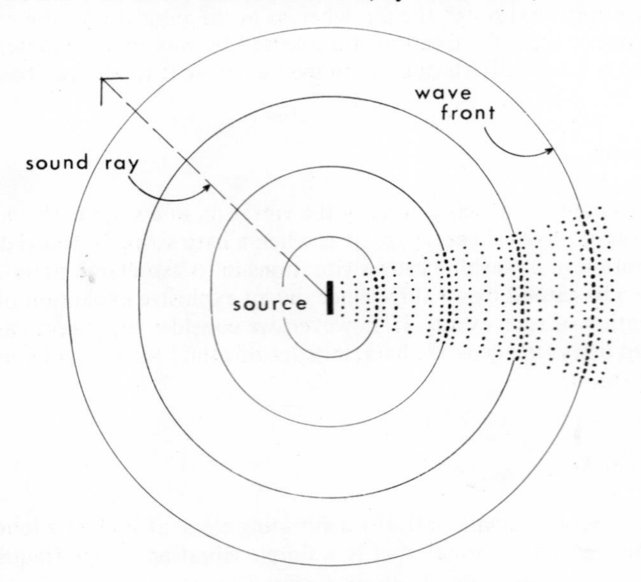

Figure 1.2

Near the surface the shape of the sound wave will be as shown in figure 1.2 but as the pressure waves expand they will become virtually spherical. It is this spherical expansion which causes a rapid reduction in sound intensity as distance from the source increases.

Near a large vibrating surface sound waves approximate to plane waves, expansion is small and therefore the rate of reduction in intensity is also small. A similar situation occurs if numerous closely spaced sound sources are arranged on a flat plane. In both situations we have what approximates to an 'area source' as compared with the 'point source' first considered.

Alternatively a number of closely spaced sound sources can be arranged along a straight line, forming a 'line source', and emitting cylindrical sound waves.

The rates of expansion of the frontal areas of spherical waves, cylindrical waves and plane waves obviously reduce in that order, affecting as we shall see the rate of reduction in sound intensity as distance from the source increases.

Directionality

Most of the individual sources of sound with which we will be concerned emit sound in all directions but many emit sound of greater intensity in one direction—obvious examples being the human voice and wind instruments. In effect, this means that the expanding sound waves, though spherical, may vary in intensity in different parts of their frontal area.

The directionality of sound sources is normally illustrated by 'polar diagrams'.

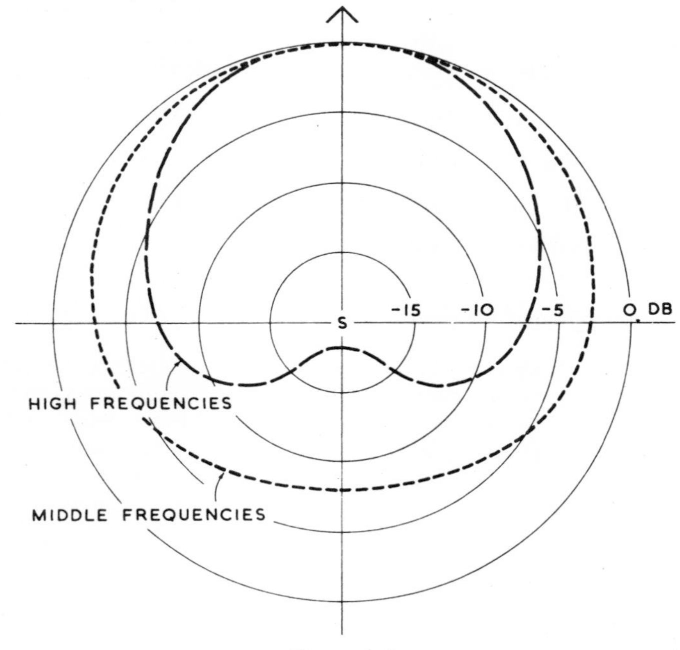

Figure 1.3

An example is given in figure 1.3 for the human voice. The speaker is facing in the direction of the arrow and the difference in intensity level, relative to direction, is shown by the curves on the graph for the high and middle frequency components of speech. It will be seen that this difference is as much as 18 dB (decibels) as between a position in front and behind the speaker—in effect much less than half as loud for the higher frequencies of speech.

Frequency and Wavelength

It has been explained that, during the passage of sound waves through the air (or any other elastic medium) the particles vibrate from side to side of their normal position of rest. The number of complete vibrations per second is referred to as the 'frequency' of the sound and corresponds with the frequency of vibration of the source—as shown in figure 1.1. One complete vibration is called a 'cycle' and therefore frequency is stated in cycles per second (or hertz).

It is frequency which determines the musical pitch of a note, sounds of high frequency having a high pitch, sounds of low frequency having a low pitch.

As will be explained below, frequency also determines 'wavelength', that is, the distance between the centres of compression of the sound waves. The wavelengths of audible sound vary from about 2 cm to about 15 m.

Figure 1.4 shows the relationship between frequency and wavelength for the fundamental tones of a piano at octave intervals. Above the keyboard are shown the frequency ranges of a few familiar musical instruments and noises. The range of frequencies of the noises would in fact be greater than shown on the drawing but the thickness of the lines is some indication of the relative intensities of sound in various parts of their frequency spectra. The frequencies of various components of the human voice can be stated approximately as follows

Male voice, vowel sounds	100 Hz
Male voice, sibilants	3000 Hz
Bass singer, bottom note	100 Hz
Soprano, top note	1200 Hz

Velocity, Wavelength and Frequency

Sound waves travel in air at about 340 m/s, irrespective of intensity or frequency. If we consider the case of a sound source generating a continuous pure tone of 100 Hz and a listener 340 m away, then the first sound wave emitted will reach the listener in 1 s during which time 99 further sound waves have been produced. In other words the intervening space will be occupied by 100 sound waves. Wavelength will therefore be

$$\frac{340}{100} = 3.4\,\text{m}$$

Thus the wavelength of a sound can be found by dividing its frequency into the speed of sound.

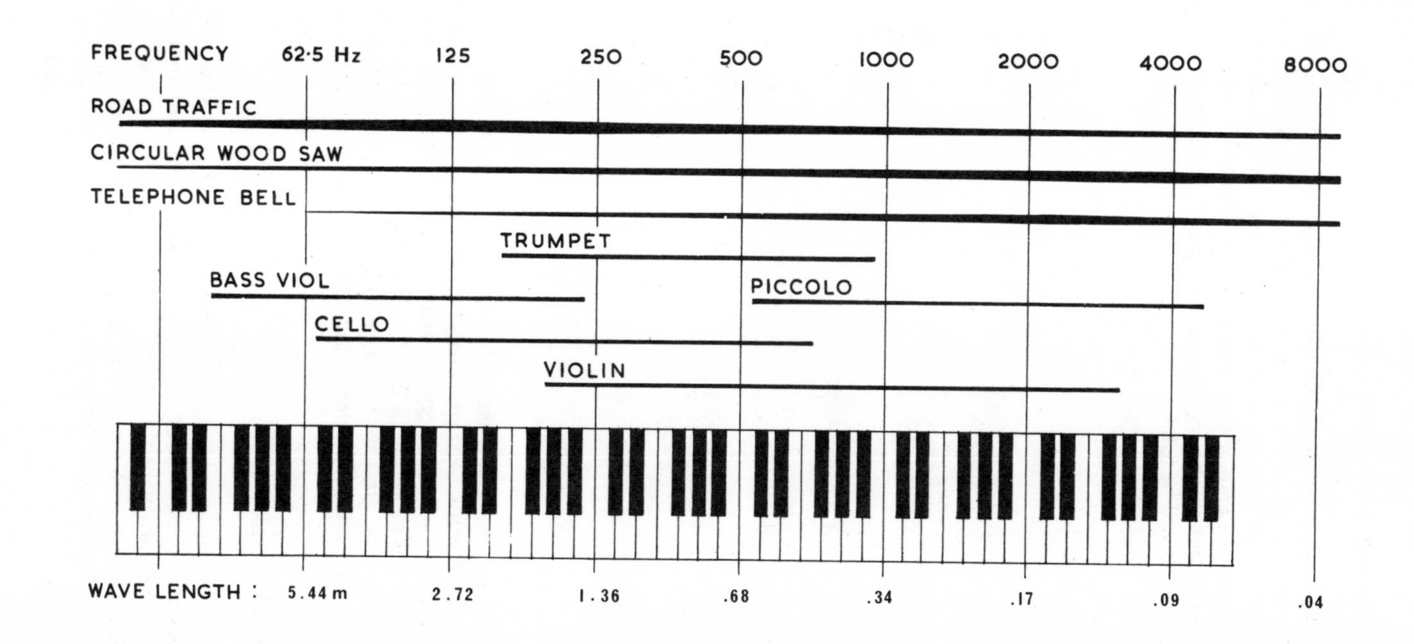
FREQUENCY
62·5 Hz
125
250
500
1000
2000
4000
8000
ROAD TRAFFIC
CIRCULAR WOOD SAW
TELEPHONE BELL
TRUMPET
BASS VIOL
PICCOLO
CELLO
VIOLIN
WAVE LENGTH :
5.44 m
2.72
1.36
.68
.34
.17
.09
.04

Figure 1.4

Sound Intensity and Loudness

If a tuning fork is struck gently the movement of its prongs will be less than if it is struck vigorously, although the frequency of vibration will remain the same. The effect on the particles of air is shown diagrammatically in figure 1.5, as is the effect on the graphical expression of amplitude of compression. The effect on the ear drum will be correspondingly greater in the second case than the first and the sound will be louder.

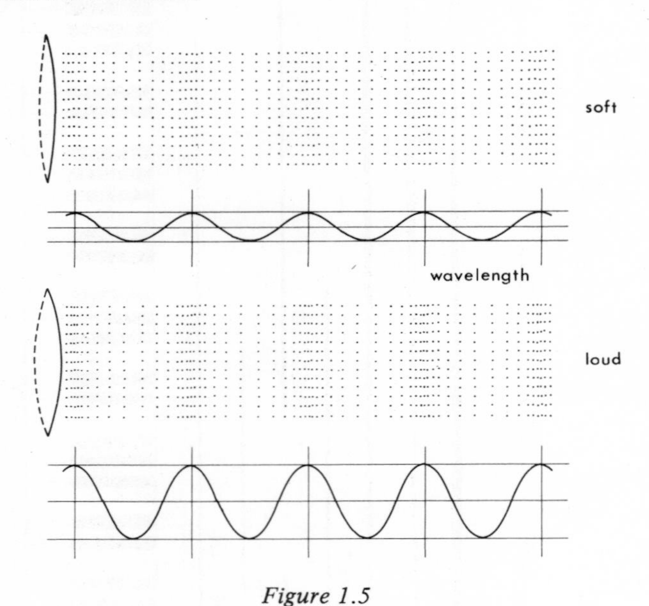

Figure 1.5

In this process sound energy is being transmitted through the air from source to listener. The total acoustic energy produced by the source is measured in watts and the flow of energy through unit area at any point in the direction of transmission is measured in W/m^2. This measure of acoustic energy is termed 'sound intensity'–relative to any given point in the sound field. For sounds of the same frequency, the greater the intensity the louder the sound.

The Decibel Scale

Sound intensity can thus be measured on a linear scale of W/m^2. It can also be measured on a logarithmic scale of decibels. To differentiate between these two forms of measurement the decibel scale is termed a measure of intensity *level.*

Most familiar scales are linear, that is, each value on the scale represents an *addition* of one unit to the preceding value, for example

0 1 2 3 4 5 6 etc. m

A logarithmic scale on the other hand is a scale of *ratios*, that is, each value on the scale represents a proportional increase on the preceding value, for example

1 10 100 1000 10000 100000 1000000 etc.

or

1 10 10^2 10^3 10^4 10^5 10^6 etc.

or as a logarithmic scale to the base 10

0 1 2 3 4 5 6 etc.

Such a scale requires a 'reference value' on which subsequent values can build by successive proportional increases. In the above scale the reference value is 1 but note that this becomes zero on the logarithmic scale. On a linear scale zero has no value.

On a logarithmic scale any value can be chosen as a reference value—to suit what is being measured. Note also that a very wide range of values can be stated by the use of small numbers.

Now, sound intensity measured in W/m^2 varies by a factor of one billion as between the weakest sound which can be heard and a sound which may be termed 'deafening'. For most purposes this form of measurement is unnecessarily fine and of course involves the use of unwieldy numbers. The logarithmic scale of decibels is therefore used and reduces the number of steps from 'threshold of hearing' to 'deafening' to about 120.

The internationally agreed reference value (RV) is 10^{-12} W/m^2—approximately the threshold of hearing for sounds of 1000 Hz. Taking this as our reference value we can now set out the scale as follows

Scale of intensity		RV							
Ratios of reference value	10^{-1}	1	10	10^2	10^3	10^4	10^5	10^6	etc.
Bels	–1	0	1	2	3	4	5	6	
Decibels	–10	0	10	20	30	40	50	60	

It will be noted that simply by using the powers of 10 we in fact arrive at a scale of *bels*. As these steps are too wide for practical purposes, the scale is subdivided by 10 to give a scale of *decibels*.

It will also be seen that, just as we can build on the reference value by proportional increases, we can also express even smaller values by proportional decreases so that –10 decibels (dB) has a value in terms of acoustic intensity, albeit extremely small and quite inaudible.

Another advantage of the decibel scale is that it corresponds approximately to the way in which we hear. For a sound to seem just perceptibly louder than another it must be increased in intensity by a certain *proportion*—in the case of pure tones under optimum conditions, by about 1 dB. Similarly, for any sound to seem about 'twice as loud' as another there must be an increase of 10 dB—and this applies to weak sounds and loud sounds alike.

The above general description of the decibel scale can now be expressed in mathematical terms as follows

$$\text{intensity level of a given sound } I \text{ relative to reference value } I_0 = \log \frac{I}{I_0} \text{ bels}$$

$$= 10 \log \frac{I}{I_0} \text{ decibels}$$

or

$$\text{difference in intensity level between two sounds } I_1 \text{ and } I_2 = 10 \log \frac{I_1}{I_2} \text{ decibels}$$

Typical Sound Intensity Levels

Some typical sound intensity levels are listed below. In looking at the figures it should be borne in mind that for sounds of similar frequency spectra an increase by 10 dB is approximately twice as loud. Since, however, the ear is more sensitive to high frequencies than low, this will not hold good for all the sounds listed. In section 2 alternative scales of subjective *loudness* will be described. The scale of decibels is an objective measure of sound energy.

Four-engine jet aircraft at 100 m	120 dB
Riveting of steel plate at 10 m	105 dB
Pneumatic drill at 10 m	90 dB
Circular wood saw at 10 m	80 dB
Heavy road traffic at 10 m	75 dB
Telephone bell at 10 m	65 dB
Male speech, average, at 10 m	50 dB
Whisper at 10 m	25 dB
Threshold of hearing, 1000 Hz	0 dB

Sound and Distance

That sound reduces with distance is common experience but, in order to predict this reduction in any given situation, it is necessary first to consider the nature of the source and the consequent form of the sound waves.

It has been seen (p. 5) that there are theoretically three types of sound source: point source, line source and area source. The rate of reduction in sound intensity will be different in each case.

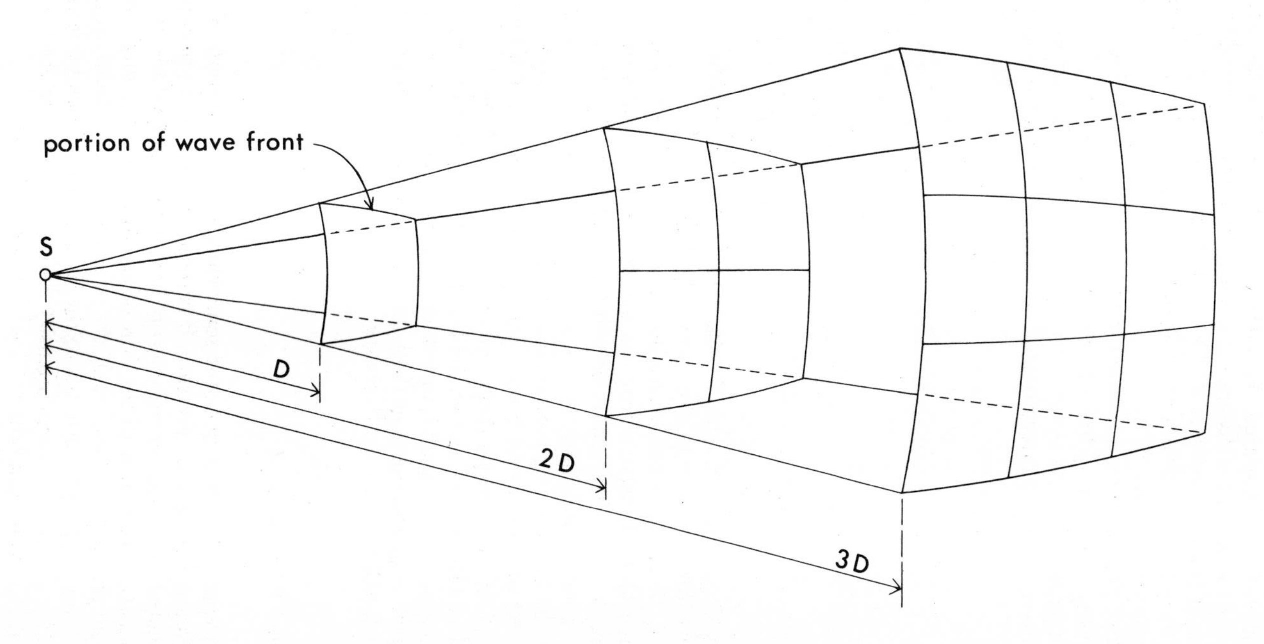
portion of wave front
S
D
2D
3D

Figure 1.6

If we first consider a point source producing spherical sound waves, figure 1.6 illustrates the proportional increase in the frontal area of the waves at three distances from the source. When the distance is doubled the area is increased by a factor of 4; when the distance is trebled the area is increased by a factor of 9, and so on. It follows therefore that the sound intensity in any part of the sound wave will be reduced by a factor of 1/4 and 1/9 respectively, as compared with the intensity at distance D.

This phenomenon is referred to as the 'inverse square law' by which sound intensity varies inversely as the square of the distance.

Since in figure 1.6 the intensity at distance D is four times that at $2D$, we can compare the sound intensity levels by using the formula on p. 10

Difference in intensity level between the two sounds I_1 and I_2

$$= 10 \log \frac{I_1}{I_2} \text{ dB}$$

$$= 10 \log \frac{4}{1} \text{ dB}$$

$$= 6.021 \text{ dB}$$

Thus each time we double the distance from the source there is a reduction in sound intensity level of almost exactly 6 dB.

In the case of a line source the expanding sound waves are, however, cylindrical and, each time distance from the source is doubled, intensity will be reduced by a factor of 1/2. The reduction will therefore be 3 dB. Figure 1.7 shows the operation of these laws, by which the *rate* of reduction in sound level progressively reduces as distance increases. The graph shows the reduction with distance in respect of a point source and a line source. In the case of a theoretical area source of infinite dimensions there is no reduction with distance.

In all the above cases we are considering the effect of distance in isolation and assuming that the sound is produced in a 'free field', that is, away from any reflecting or absorbent surfaces. In practice, other factors cause sound reduction, which will be discussed in section 3.

Sound Pressure Level

As explained on p. 3 air-borne sonic vibrations cause variations in pressure, plus or minus atmospheric pressure. It is these fluctuations in pressure which are detected by the ear, or the microphone of a sound level meter. Whereas sound intensity is a measure of the flow of acoustic energy through the medium, sound pressure is a measure of the resulting *variation* from normal atmospheric pressure—and is proportional to the square root of sound intensity.

Just as *ratios* of sound intensity can be expressed on the logarithmic scale of decibels, so also can ratios of sound pressure. Thus

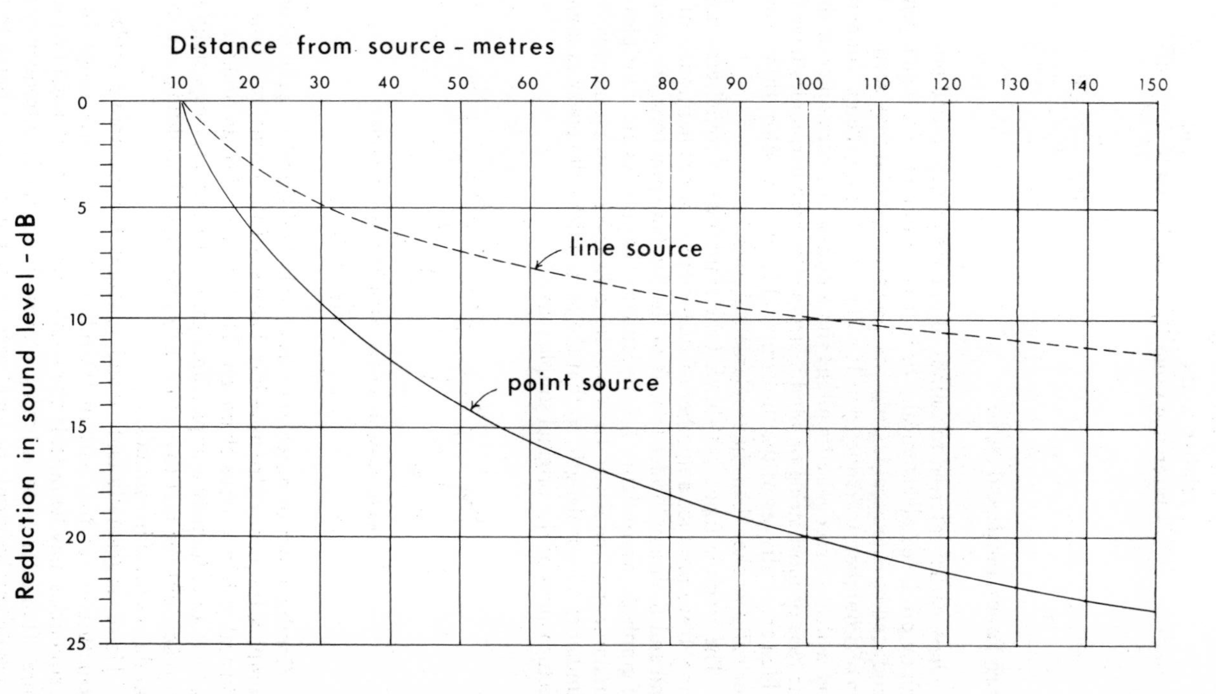
Distance from source - metres
10
20
30
40
50
60
70
80
90
100
110
120
130
140
150
Reduction in sound level - dB
0
5
10
15
20
25
line source
point source

Figure 1.7

sound pressure level of a given sound P relative to the reference value P_0

$$= 10 \log \frac{(P)^2}{(P_0)^2} \text{ dB}$$

$$= 20 \log \frac{P}{P_0} \text{ dB}$$

Pure Tones, Music and Noise

We have earlier referred to sounds of one frequency, called pure tones, such as can be produced electronically or by a good tuning fork struck gently. A note produced by a musical instrument, however 'pure' it may seem to the ear, is made up of a number of frequencies. It will have what is called a 'fundamental tone' (by which we recognise its pitch) and a number of harmonics or overtones. In the case of most string or wind instruments these harmonics have frequencies which are multiples of the fundamental frequency. Also, these harmonics are normally weaker than the fundamental tone.

The graph in figure 1.1 showed the amplitude of sound pressure for a pure tone. If we produce a number of such graphs for the harmonics of a musical note and add their amplitudes to that of the fundamental tone, we produce the kind of graph shown in figure 1.8. Because the harmonics are multiples of the fundamental frequency, their wavelengths will be simple subdivisions of the fundamental wavelength, so that the harmonics will appear as minor but regular disturbances of the original simple curve.

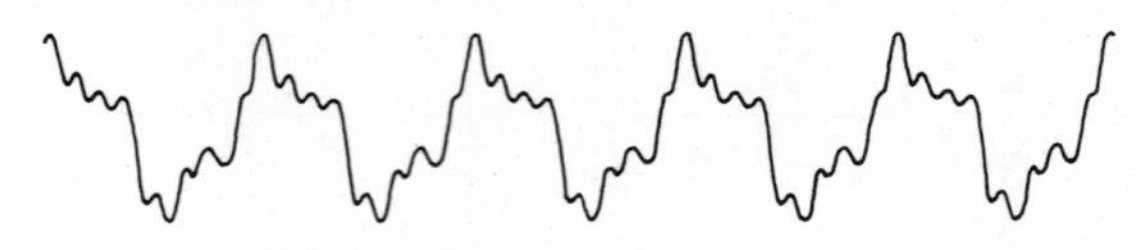

WAVE FORM OF A NOTE PRODUCED BY A CLARINET

Figure 1.8

Combinations of notes, such as chords, will therefore present an even more complex frequency structure but, because of their tonal relationship retain a musical or harmonious character. The total sound of an orchestra can thus be seen as an extremely complex amalgam of frequencies, especially when one takes into account the fact that some instruments, such as drums and cymbals, do not produce simple harmonics.

Noise, as generally understood, is sound of even more complex frequency structure, of random pattern extending over a very wide range. In fact some frequencies may lie outside the audible range. Typical machine noise is therefore often referred to as having a 'broad-band, random-frequency structure'. Lacking any harmonic or tonal pattern such noise seems harsh to the ear. Nevertheless the dominant frequencies can often be recognised—as when we say that a noise is 'shrill' or 'booming'. In other words, certain groups of frequencies have more power than others.

Addition of Two or More Sounds

When two sounds occur simultaneously the resulting intensity level cannot be obtained by the addition of the sound levels of each measured separately—any more than one would add the first, second, third and fourth floor levels of a building to arrive at a total height of ten storeys.

Apart from the fact that we are using a logarithmic scale of values and such values cannot be added arithmetically, the combination of two sounds will also depend on the nature of the sounds and whether, in the case of pure tones, they are produced in or out of phase. If two pure tones of equal intensity are produced 'in phase', that is, in such a way that the centres of compression of their sound waves exactly coincide, then there will be an increase of 6 dB compared with one of the sounds measured separately.

In the case of two noises of wide frequency range and random frequency structure, their combination, if each is of equal intensity level, results in an increase of 3 dB. If they are not of equal intensity level then the increase will be less. Similarly, the increase in sound level due to the combination of a number of noises is not proportional to their number. As each noise is added in the increase in total noise tends to diminish. In section 3 on p. 109 is given a scale for the addition of noise levels.

Frequency Spectrum

Any noise can by analysed with suitable instruments to find the sound intensity level in various parts of its frequency range. Normally such an analysis is made in octave or one-third octave bands, the sound level being recorded for each group of frequencies. Examples of such analyses are shown in figure 1.9, illustrating the 'sound spectrum' for each noise.

From the examples shown it will be seen that in the case of the circular saw there is more sound energy in the high frequencies than the low, as would be

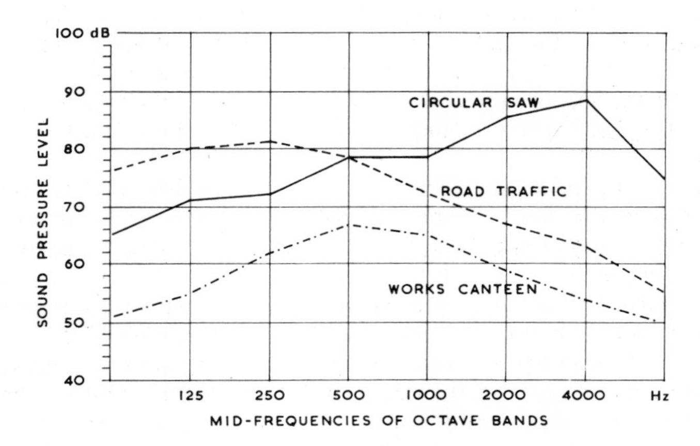

Figure 1.9

expected from the shrill character of this type of noise. The 'roar' of traffic on the other hand is indicated by the shape of the graph for traffic noise.

The sound intensity level of any of these noises can be expressed as one value. In the case of the circular saw it would be about 91 dB, this being the over-all sound level as registered by a simple sound level meter.

Sound Transmission in Solid Materials

We have so far considered sound energy transmitted through the air. Sonic vibrations can also travel through liquids and solid materials, of which building structures will be our main concern.

There are two kinds of sound transmission by way of solid materials: firstly by particle vibration (essentially similar to air transmission) and secondly by panel vibration. The first involves vibrations within the material itself and the second involves the vibration or resonance of the material as a whole. The following examples will illustrate both forms of transmission.

If a tuning fork is struck and the shaft held against the end of a long steel rod, its vibrations will pass down the length of the rod by the oscillations of the particles of steel. The wave motion is similar to that of sound in air, except that its speed and absorption rate are different. Also the energy of vibration is almost entirely confined within the solid material. When the sound waves reach the end of the rod they are reflected back; very little energy will leave the rod and be conveyed to the surrounding air. The reflections will pass to and fro along the length of the rod until their energy is absorbed by particle friction and dissipated in minute quantities of heat.

If the ear is placed in direct physical contact with the end of the rod, the sound of the tuning fork can be clearly heard, even if the rod is very long and passes into an adjoining room. The sound can more easily pass from one solid material (the rod) to another (the ear), than from a solid material into the air. If the ear is positioned quite close to, but not in contact with, the end of the rod, nothing will be heard. The experiment is illustrated in figure 1.10.

If, however, we fit a wood panel on the end of the rod we find that we can now hear the sound of the tuning fork quite clearly without putting the ear in contact with the panel. The surface area of the panel is sufficient to impel air-borne sound waves as the panel is moved to and fro by the vibrations travelling down the length of the rod. The panel is said to 'amplify' the sound in a similar manner to a loudspeaker cone.

Finally, if along the length of the rod a section of resilient material is inserted (such as a few centimetres of rubber tube) this will absorb, or damp out, the vibrations as they pass along the rod and audible sound will no longer be emitted by the panel.

All these phenomena may be present in a building when a vibrating element, such as a machine, is mounted on the structure. The vibrations of the machine can be carried through the structure for considerable distances and light panels, such as partitions, can convert the vibrations into air-borne sound in other parts of the building. Any discontinuities in the structure will, however, impede the transmission of the vibrations, as will the presence of any resilient material along their path—for example, if used in the mounting of the machine.

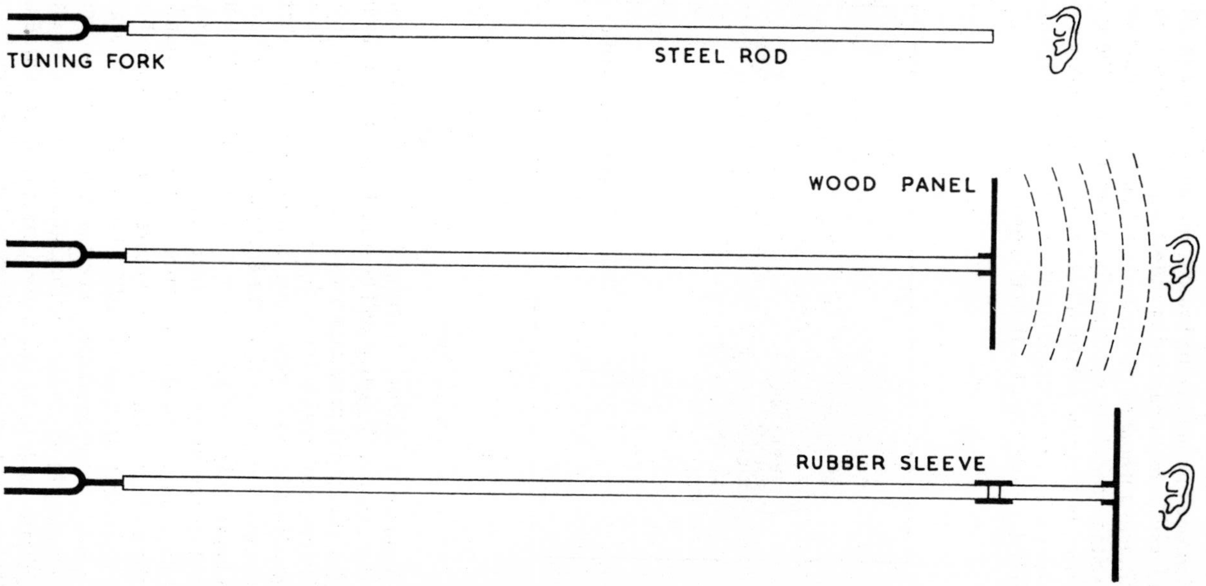

Figure 1.10

Panel Transmission

The second category of sound transmission 'through' a solid material is the more familiar one of sound being heard from one side of a partition to the other. Here, sound waves are propagated in the air adjoining the partition and cause the partition *as a whole* to vibrate and emit air-borne sound waves on the other side. The sound waves have not, literally, passed through the solid material of the partition. In fact, the partition has become a secondary source of sound. The situation is illustrated diagrammatically in figure 1.11, although the vibration of the panel is more complex than is shown.

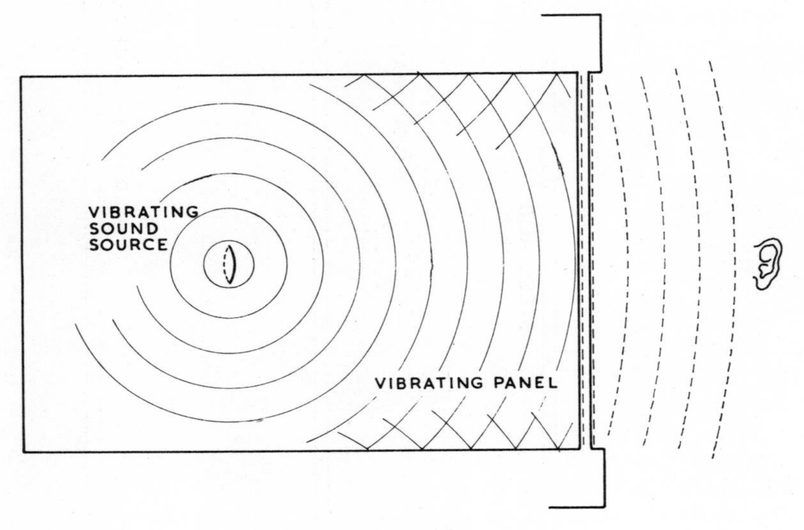

Figure 1.11

A number of factors affecting this kind of sound transmission can be illustrated by this example. Firstly it will be noted that the panel will be vibrated not only by the alternating pressure waves of direct sound but also by reflected and inter-reflected sound within the enclosure, thus adding to the amount of sound energy transmitted by way of the panel. It follows therefore that the sound level external to the panel will be lower if the enclosure is lined with sound-absorbing material, not because the insulation value of the panel has increased, but because the sound energy level within the enclosure has been reduced. If, on the other hand, the direction of sound propagation is reversed, the sound source being external to the enclosure, lining the enclosure with sound-absorbing material will reduce the reverberation of sound radiated by the panel and again the panel will *appear* to provide better insulation, whereas its insulation properties have not changed.

The way in which the reverberant sound within the enclosure causes the panel to vibrate needs to be understood, at least in general terms, if the considerable differences in insulation provided by building panels is to be explained.

Since the panel is being set into vibration by the fluctuating pressure waves of air-borne sound, it is to be expected that the weight of the material per unit area will influence the amplitude of vibration, since the pressure waves have to overcome the inertia of the panel. This is generally true and gives rise to the 'mass law' by which

(1) sound transmission reduces as the weight per unit area of the panel is increased, and

(2) sound transmission reduces as the frequency of incident sound rises.

However, factors other than mass affect the behaviour of the panel, causing considerable deviation, in the case of rigid panels, from what might be expected by the operation of the mass law alone.

At very low frequencies the stiffness of the panel, and therefore its resistance to deformation, may have more effect than its weight, so that we find that the first variation from the simple mass law occurs. In this part of the frequency range insulation is said to be *stiffness controlled.*

If a rigid panel is struck it will continue to vibrate at frequencies which are determined by its size, shape and thickness. We may call this its 'natural mode of vibration' and the frequencies at which this occurs, its 'resonant frequencies'. When incident sound waves have frequencies similar to these resonant frequencies vibration will be increased and transmission will be greater. Insulation will then be reduced below what would be predicted by the mass law. At these frequencies insulation is said to be *resonance controlled.*

The most important deviation from a steady increase in insulation as frequency rises is, however, due to what is called the 'coincidence effect'. As will have been seen in figure 1.11 sound transmission by way of a panel will not only be caused by sound waves impinging on it at right angles to its surface but also by those striking it obliquely, as shown in figure 1.12. These will produce a forced motion in the panel which has a greater wavelength (called the 'trace wavelength') than that of the incident waves in air. At the same time, what are termed 'free bending waves' can be propagated in the panel which, at a frequency dependent on the nature of the panel, can match the trace wavelength of the forced vibration. When this occurs transmission is increased and insulation is reduced. This phenomenon is called the 'coincidence effect'.

The frequency at which this occurs is called the critical frequency. In the region of this frequency sound transmission is said to be *coincidence controlled.*

This critical frequency varies considerably, depending on the density, thickness and stiffness of the panel. For a single brick wall it will be in the region of 100 Hz, whereas for a 10 mm plywood panel it will be approximately 1300 Hz.

Panel Transmission and Frequency

It will be seen from the above that the degree of sound transmission 'through' a panel, and thus its sound insulation value, will vary with frequency owing to the interaction of a number of factors which influence the amplitude of its vibration in different parts of the frequency spectrum. Whereas there is a general rise in insulation as the incident sound frequency increases, there may be serious weaknesses in insulation to be taken into account. These are summarised by the graph in figure 1.13, which indicates the sequence and general effect of the phenomena

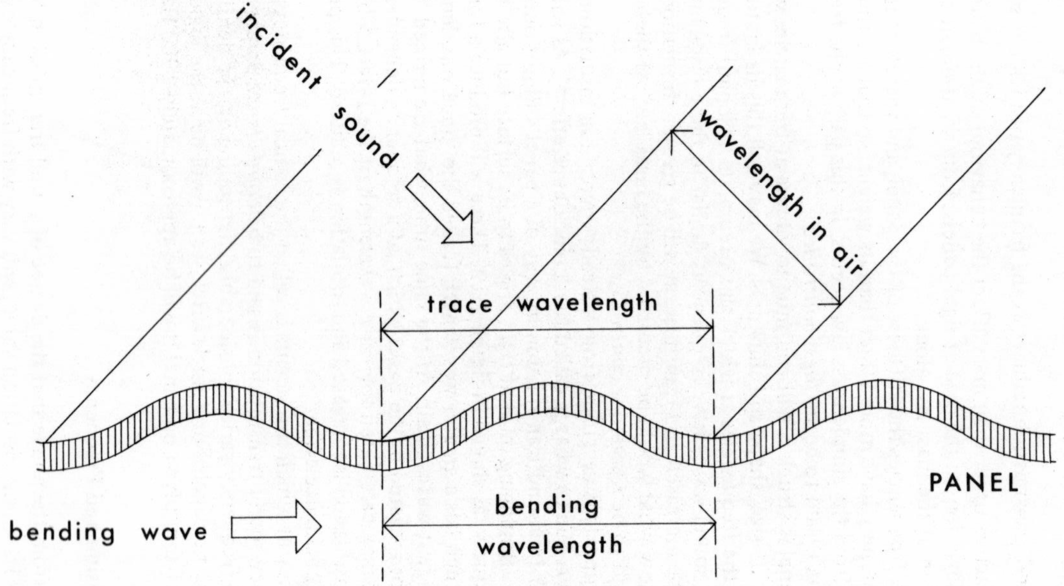

incident sound

wavelength in air

trace wavelength

PANEL

bending wave

bending wavelength

Figure 1.12

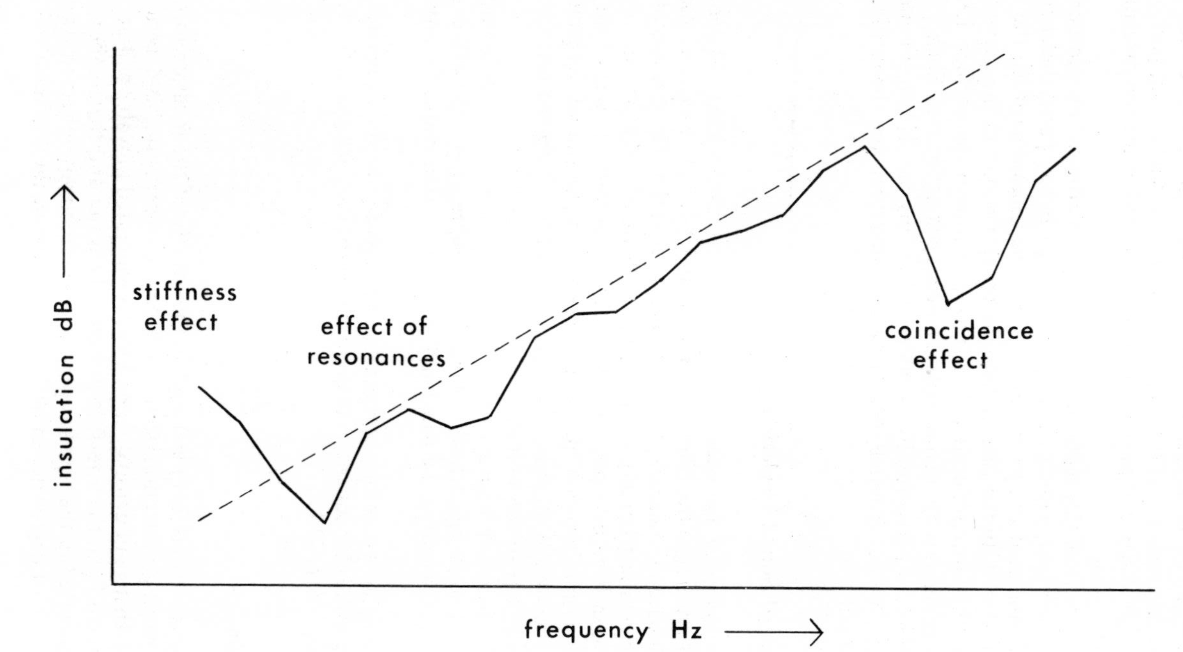
insulation dB
stiffness effect
effect of resonances
coincidence effect
frequency Hz

Figure 1.13

described above. The dotted line shows the rise in insulation that would be expected by the operation of the mass law alone. No values have been put along the two co-ordinates of the graph since, as explained above, the 'dips' in the curve of insulation can occur in almost any part of the frequency spectrum.

As will be seen in section 3, if the coincidence effect occurs in that part of the frequency range which is important subjectively sound insulation measures may be invalidated.

However, the *extent* of this reduction in insulation is affected by another characteristic of the panel, namely, 'internal damping'. In any panel the energy of vibration is partly radiated and partly absorbed by the friction between the particles of the material. In the case of some materials the dips in the curve shown in figure 1.13 are therefore reduced and are not so serious in their effect. Lead is an example of a material with high internal damping characteristics. Such a material is, however, not structurally suitable for building panels but, if combined with a more rigid material, internal damping can be increased and the insulation curve 'smoothed out'.

The Measurement of Sound Insulation

The term 'insulation value' has been used above to describe, as a general concept, the resistance of a panel to the transmission of sound.* It is, however, necessary to be able to evaluate, or measure, this resistance so that the effectiveness of an enclosure can be estimated in any given situation. Obviously, the amount of sound energy transmitted by a panel will vary according to the power of the sound source. The *ratio* between the sound intensity on one side of the panel and the sound intensity on the other is, however, constant for any given panel, regardless of the power of the source.

Since the decibel scale is a scale of *ratios* of sound intensity, the difference between the sound levels on each side of the panel (expressed in dB) will express the insulation effectiveness of the panel and will be a constant value for any given panel.

Insulation values are therefore expressed in decibels at stated frequencies, or averaged over a stated frequency range. In this form they are given, for typical building elements, on p. 111.

Sound Reflection

Whenever sound vibrations meet a change in the density of the medium through which they are passing, reflection occurs. The degree of reflection is dependent on the degree of change in density. Sound waves generated in water will be effectively reflected from the interface between the water and the air above. Sound waves generated in the air of a room will be effectively reflected from its bounding surfaces to a degree dependent on the nature of the surface.

*Alternative terms for 'sound insulation value' are 'sound transmission loss' and 'sound reduction index'. To avoid confusion only the one term 'sound insulation value' will be used in this book.

In the subject of room acoustics we shall be mainly concerned with the latter situation. Unless such surfaces are treated with porous material, most of the sound energy generated in the room will be reflected. By definition, the percentage of sound energy reflected is the reciprocal of the percentage absorbed. Reference therefore to the coefficients of absorption for various materials, given on pp. 198 to 202, will indicate the degree of reflectivity. It should also be noted that absorption, and therefore reflectivity, varies with frequency.

After reflection the intensity of the sound will, however, be affected by the *shape* of the surface as well as its absorption characteristics. To understand this we need to understand the geometry of sound reflection and its effect on the intensity of sound waves.

Reflections from Plane Surfaces

Figure 1.14 illustrates the geometry of sound reflections from a flat, smooth surface. The reflected sound waves are spherical and their centre of curvature is the 'image' of the source of sound. The image is on a line normal to the surface and at the same distance from the surface as the source.

It will be seen therefore that the reflected sound will attenuate *at the same rate* as sound in a free field, that is, in accordance with the inverse square law (see p. 10). The intensity of the reflected sound waves will therefore depend on their distance from the image and the degree of absorption occurring at the reflecting surface.

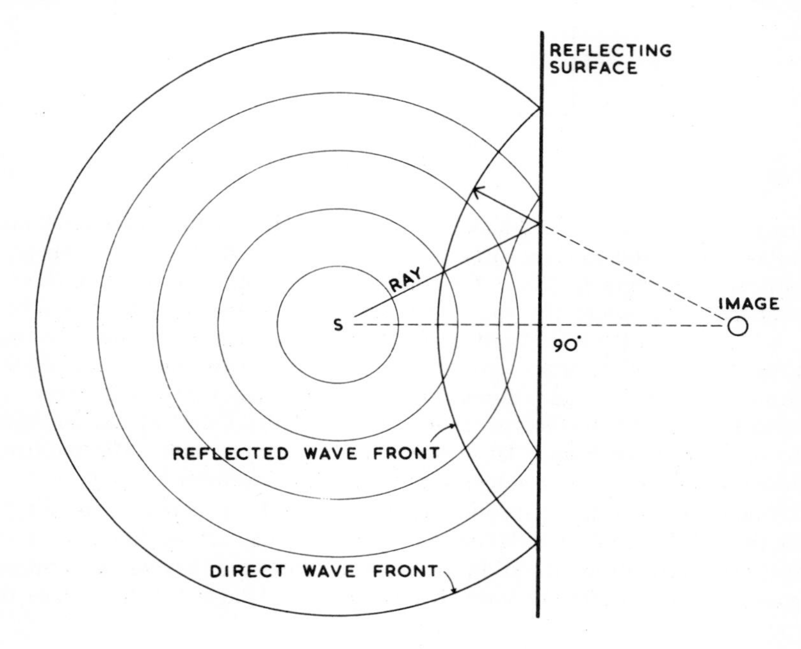

Figure 1.14

Reflections from Curved Surfaces

The geometry of reflections from curved surfaces is best derived from the employment of sound 'rays'. An example of a sound ray was shown in figure 1.14 and it can be defined as the *direction* of propagation of the sound wave.

The first drawing in figure 1.15 shows that a reflected sound ray is on a radial from the image in the case of a flat surface. The angle of reflection of the ray is equal to the angle of incidence to the surface. The second drawing in figure 1.15 shows that rays striking a curved surface are each reflected so that the angle of reflection is equal to the angle of incidence to radials drawn at their points of contact. Each ray will in effect have its own image; the wave front will not be part of a circle and must be found by drawing each ray of equal *total* length as shown.

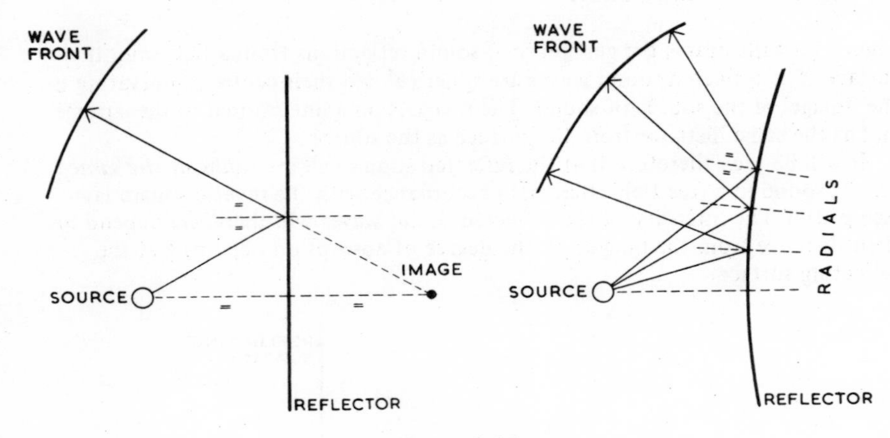

Figure 1.15

Figures 1.16, 1.17 and 1.18 make a direct comparison between the reflections from flat, convex and concave surfaces. The distance from the source to the reflector is the same in each case, the cone of sound considered is the same, and the time interval at which the wave fronts are drawn is also the same. It will be seen, however, that the wave front from the convex surface is considerably bigger than that from the flat surface, and that the wave front from the concave surface is considerably smaller, and is diminishing. It follows then that sound waves reflected from convex surfaces are more attenuated, and therefore weaker, than sound waves reflected from a flat surface. Similarly, sound waves reflected from a concave surface are more condensed and therefore of greater intensity.

It should also be noted that sound waves reflected from a concave surface, unlike those reflected from a flat or convex surface, may actually increase in intensity the further they travel. In the example illustrated they will pass through a region of focus in which the sound heard may be as loud as that heard near the source.

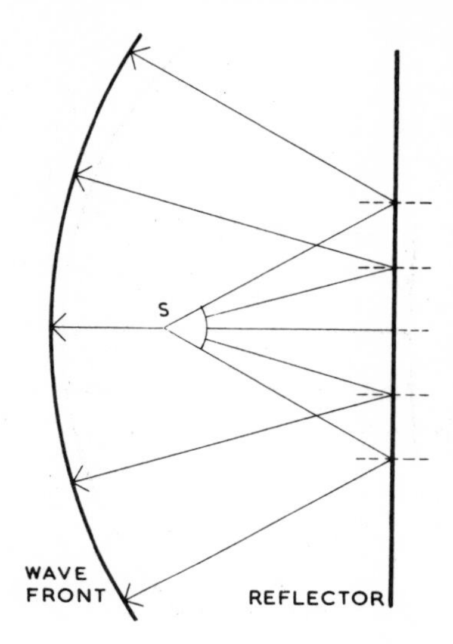

Figure 1.16

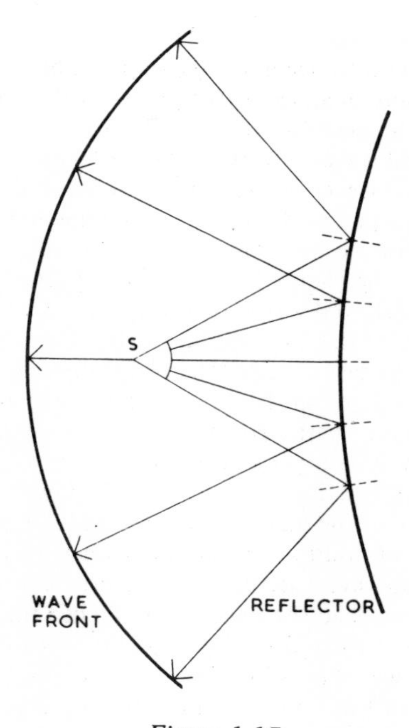

Figure 1.17

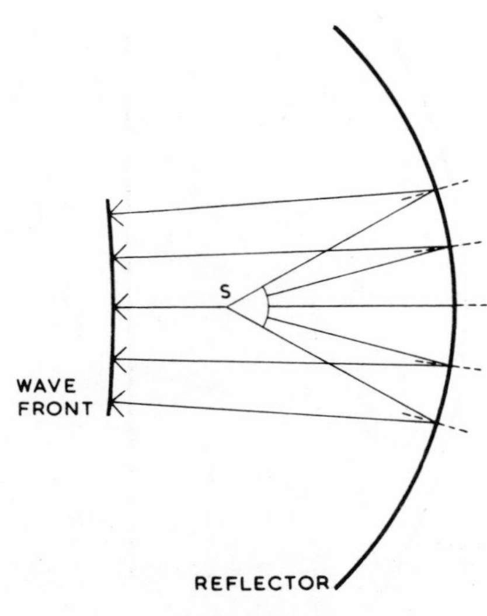

Figure 1.18

Reflectors of Limited Size

For effective sound reflection to occur a reflector must be large in relation to the wavelength of the sound and, in all cases, the power of the reflected sound will be affected by diffraction at the edges of the reflector.

Figure 1.19 shows sound waves reflected from reflectors of different widths together with the diffracted waves (or wave fringes) which develop from the edges of the reflectors. The frequency of the sound is the same in both cases, as indicated by wavelength, and therefore the degree of diffraction is also similar.

However, in both cases, the energy in the diffracted sound waves is being extracted from the main reflected waves and, because the latter are smaller in the second example, the effect of this loss will be greater. For a given frequency therefore small reflectors are less efficient than large ones. When reflectors are used in auditoria to reinforce sound they must therefore be of adequate size.

'Edge diffraction', however, varies considerably with frequency. Figure 1.20 compares reflections from reflectors of the same size but where the sound is of low and high frequency. Diffraction in the case of the lower frequency is greater and therefore the intensity of the main reflected waves is less. It follows therefore that in the case of sounds of multiple frequency, such as speech or music, a reflector of limited size can have a selective characteristic, projecting the higher frequencies more effectively than the lower ones. To put this another way, it may be said that such reflectors have a 'directional quality' for the higher frequencies which they lack for the lower. This should be borne in mind when interpreting the diagrams in section 4 of this book, and indicates a certain limitation on the validity of the geometric analysis of sound propagation in an auditorium.

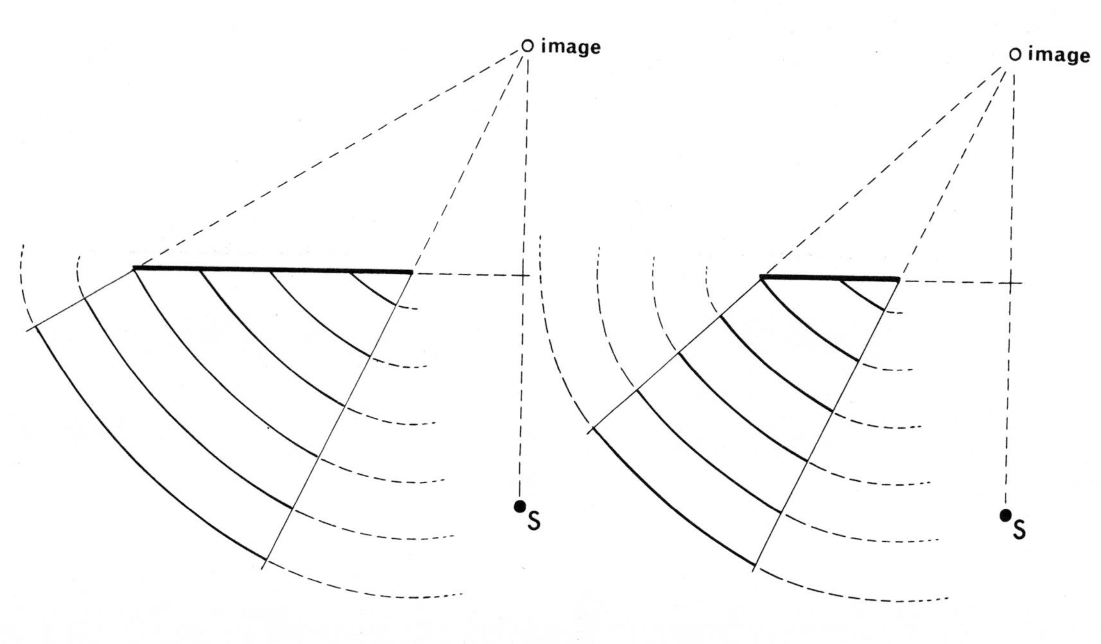

Figure 1.19

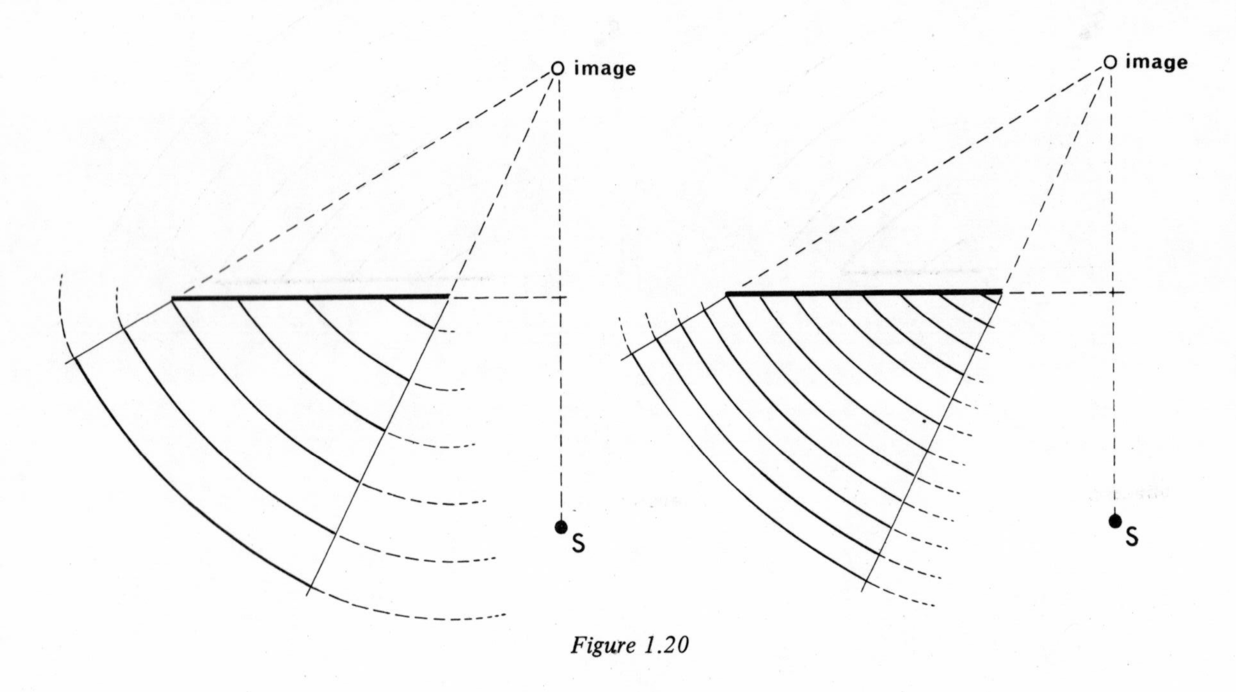

Figure 1.20

Nevertheless, since the intelligibility of speech is much more dependent on hearing the middle and high frequencies than on the reception of low frequencies, reflectors of about five times the wavelength of middle frequencies (about 3 m) are very effective in reinforcing speech sounds.

Dispersion

Sound waves impinging on a modelled surface will, on reflection, be broken up into a number of small and weak waves, as shown in figure 1.21, provided the modelling of the surface is sufficiently bold in relation to wavelength. This phenomenon is referred to as 'dispersion' or 'scattering'. The distance between the changes of direction or 'breaks' in the surface must be at least one-tenth of the wavelength for such frequencies to be effectively scattered. Small modelling or texturing of the surface will result in normal reflection for all but the highest frequencies. In auditoria, surfaces required to disperse sound reflections are usually modelled with surface breaks of the order of 50 cm.

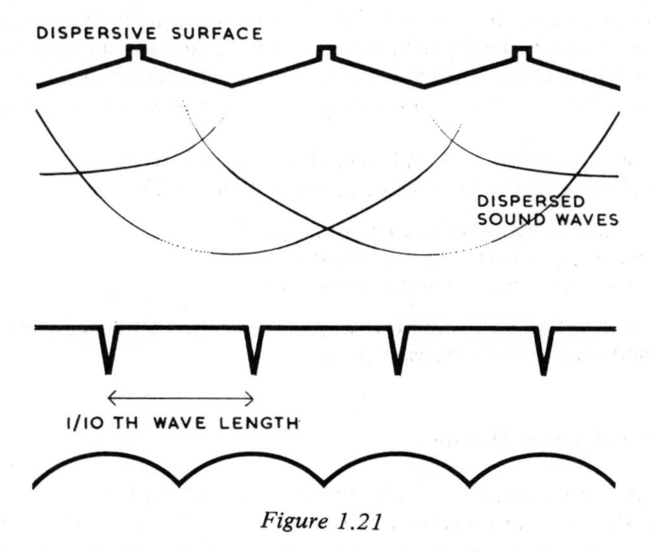

Figure 1.21

The result of such breaking up of the reflected sound waves is to reduce the intensity of the reflected sound, so that modelled surfaces are often employed to reduce the power of echoes, to prevent standing wave patterns from forming between parallel opposing surfaces or to assist in providing diffuse reverberation.

Reflections from Re-entrant Angles

Sound entering a right-angled corner of a room will be reflected back towards the source if the adjacent surfaces are of a reflective material. This is shown in figure 1.22 (A). Such reflections are often the cause of disturbing echoes, as will

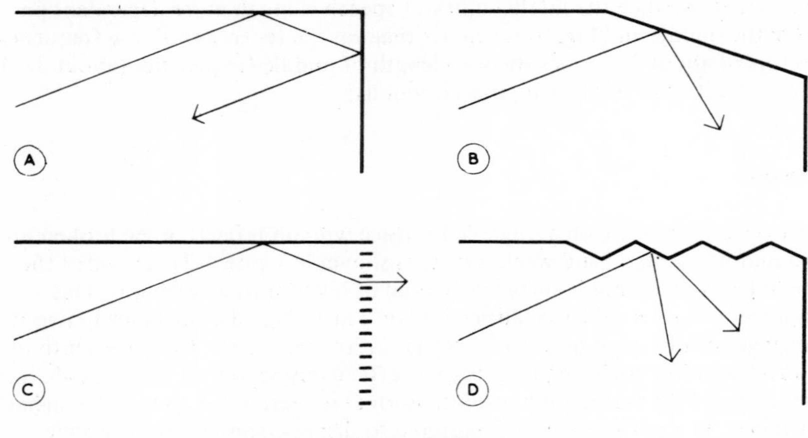

Figure 1.22

be shown in section 4. As in the case of all reflections, the phenomenon is frequency dependent, that is, related to wavelength and the dimensions of the reflecting surfaces. Quite small areas of reflecting material in the corner of a room, for example between ceiling and wall, can, however, result in high-frequency echoes.

To prevent this return of sound towards the source the corner can, however, be modified in any of the three ways shown in figure 1.22

(B) it may be made other than a right-angle
(C) one surface may be made absorbent, or
(D) one surface may be made dispersive.

Absorbent or dispersive treatment, if employed for this purpose, must, however, be taken right into the corner, as shown.

Diffraction and Sound Shadows

When sound is interrupted by an obstruction a 'sound shadow' can be formed behind it if the obstruction is large compared with the wavelength of the sound. Diffraction, however, occurs at the edge of the obstruction, causing sound to enter the 'shadow'. The degree of penetration into the shadow is frequency dependent, high frequencies being less diffracted than low frequencies. A screen will nevertheless *reduce* the level of sound for a listener behind it and can be employed, for example, as a defence against road traffic noise. In auditoria, small obstructions such as piers will have no effect on the reception of sound behind them but, as will be seen in section 4, large elements such as balconies can present acoustic problems.

Figure 1.23 shows the diffraction of sound waves passing over the edge of an obstruction. The curvature of the diffracted waves is centred on the edge of the screen and they become weaker the further they enter the 'shadow'. In the case of sounds of multiple frequency (speech, music or noise) the 'coloration' or

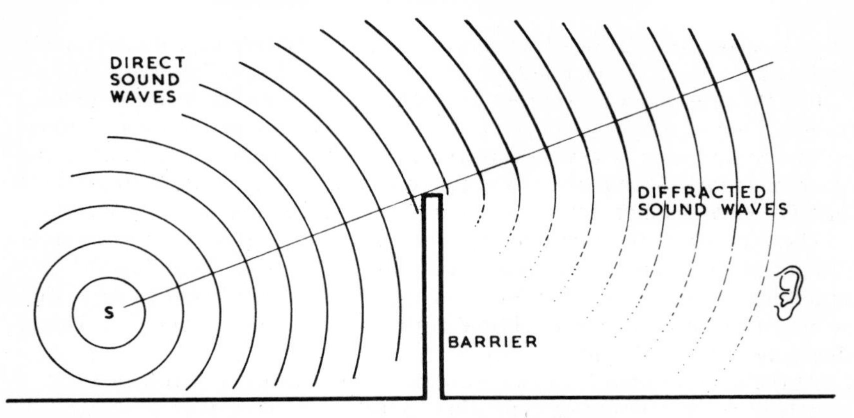

Figure 1.23

frequency spectrum of the sound will be modified by the presence of the screen because the lower frequencies will be diffracted to a greater extent than the high frequencies.

Standing Waves

Standing waves are, as the term implies, stationary fluctuations in pressure due to the superposition of sound waves moving in opposite directions. Figure 1.24 attempts to explain the phenomenon. The first graph shows a normal 'travelling

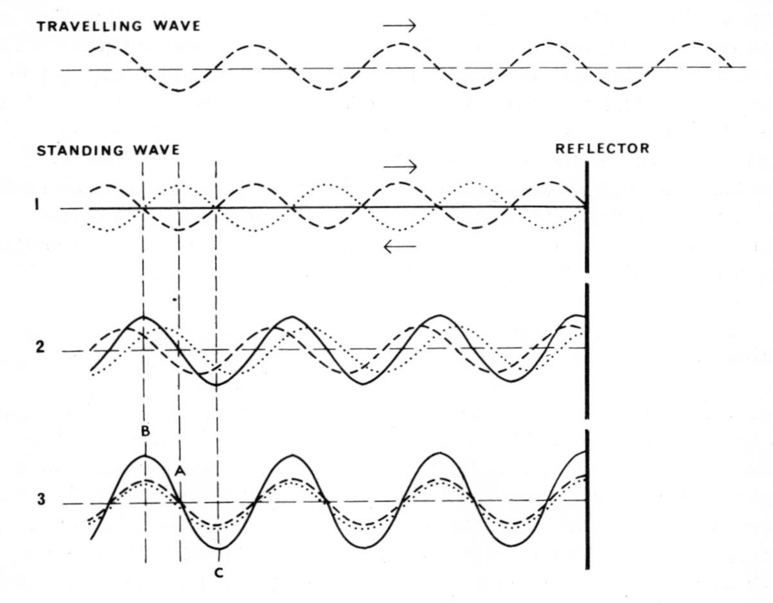

Figure 1.24

wave' in a free field. Below this is shown the effect on this wave if interrupted by a reflecting surface. Three points in time are examined (numbered 1 to 3).

In 1 the reflected wave (dotted line) is turned back on the incident wave (broken line) so that zones of greatest compression are superimposed on zones of least compression. The waves are said to be 'out of phase'. The effect is that of a momentary cancellation of sound pressure, depicted by the horizontal solid line. The arrows indicate the movement of the incident and reflected waves.

The situation at the next point in time is shown in 2 when the two waves are still out of phase but not completely cancelling each other. By adding the amplitudes of the two waves (plus or minus atmospheric pressure) a curve can be drawn which shows the amplitude of the two waves in combination. This is shown again by a solid line.

At the point in time 3 the two waves have now moved so that they are exactly in phase, reinforcing each other to the maximum extent, as shown once more by a solid line.

If we now compare the situations in all three graphs, *referring only to the solid lines*, we will see that at point A there is always an absence of compression, whereas at points B and C there will be fluctuations of pressure as the process continues. In other words, at positions A (half or multiples of half the wavelength from the reflector) air is always at normal pressure and there is an absence of sound, whereas at positions B and C there are variations of pressure causing sound, and, assuming total reflection, these variations will have twice the amplitude of the original travelling wave. Thus, depending on *where* we are listening, sound will vary in intensity.

If sound is generated between two reflecting surfaces, parallel to each other and spaced at an exact multiple of the wavelength, then by inter-reflection the process will be repeated between the reflectors. This will occur not just for one frequency (or wavelength) but for all those sounds whose wavelengths 'fit' the space between the reflectors. At each of these frequencies sound will vary in intensity, depending on the *position* of the listener. For the higher frequencies where wavelength is small the mere movement of the head will result in a fluctuation in the loudness of the sound.

For this and other reasons which will be discussed in section 4 parallel reflecting surfaces are best avoided in auditoria. In smaller rooms they can cause disturbing acoustic distortions. Either the surfaces should be made non-parallel or at least one surface should be made absorbent or dispersive.

Reverberation

Subjectively, reverberation can be defined as the continuation of audible sound in an enclosed space after the sound source has been cut off. It is a familiar phenomenon in large rooms containing few sound-absorbing materials, such as churches, where reverberation can be heard lasting five or more seconds.

When a pulse of sound is generated in an enclosed space two phenomena may occur. Sound of short wavelength relative to the dimensions of the room is reflected and inter-reflected in a manner which is diagrammatically illustrated in

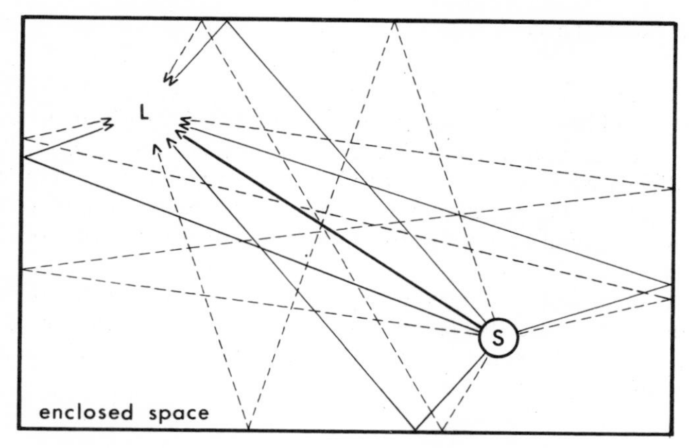

Figure 1.25

figure 1.25. Superimposed on this random scattering of sound will be, in the case of rooms having parallel reflecting surfaces, a standing wave pattern of those sounds which have wavelengths related to the dimensions of the room.

If we first consider the random inter-reflection of sound illustrated in figure 1.25 it will be seen that a pulse of sound at S will be first heard by the listener (L) by direct path (the heavy solid line) followed by six primary reflections from walls, floor and ceiling (the light solid lines) with varying time delays, followed by six secondary reflections (the dotted lines) with more extended time delays. The process will continue until by attenuation of the sound waves and by absorption in the air and at bounding surfaces the reverberant sound will become inaudible.

Assuming that all six surfaces are equally reflective, it will be seen that, the later the reflections arrive at L, the weaker they will be by virtue of the fact that they have travelled further and been reduced by more contacts with bounding surfaces. In other words, the listener will hear the pulse of sound followed by a series of weakening reflections until the reverberation of the sound decays to inaudibility. Because of the normally very short time intervals between the reception of one reflection and the next, in most situations reverberation will be heard as an extension of the original sound and not as a series of discrete sounds. The situation in which a discrete sound or echo is heard will be described in section 4.

In the case of an enclosure having parallel, plane reflecting surfaces standing wave patterns will be superimposed on this general decay of reverberation for those frequencies whose wavelengths 'fit' the distance between the surfaces. This, as we have seen, will result in an increase in their intensity and therefore an extension of their decay.

The factors that affect the *duration* of reverberation, the measurement of reverberation time and its significance in the design of auditoria will be discussed in section 4.

Reverberant Sound Level

If we now consider a situation in which, not a pulse of sound, but a continuous sound is generated in the enclosure, then the listener will first hear the sound by direct path, at the level it would be heard in the open air. The sound will, however, immediately increase by the reception of a continuing series of reflected sound waves. The sound will in fact 'build up' by reverberation until a level is reached at which sound energy lost by absorption equals the output of energy at the source.

As long as the sound continues to be generated the sound in the room will remain at this level, called the 'reverberant sound level'. When the sound is cut off at the source it will begin to decay. The phenomena of 'build-up' and 'decay' are shown graphically in figure 4.21.

Sound Absorption

For our present purposes, sound absorption can conveniently be divided into two categories: absorption during the passage of sound through the air, and absorption on contact with an obstruction.

In the former case, referred to as 'air absorption', loss of sound energy is due to the friction of the oscillating molecules of air during the passage of sound waves. Sound energy is converted into minute quantities of heat energy. The loss of energy is small except over large distances, and varies with frequency. It will be appreciated, however, that, in considering the reduction in sound level with distance, this is an additional loss which has to be taken into account where distances are large. In room acoustics, due to the extended paths of inter-reflected sound (reverberation), air absorption is also significant, especially at higher frequencies.

When sound impinges on an obstruction absorption takes place to a degree that depends on the nature of the material or its surface. Absorption can be due to surface friction, friction due to resonance in any cavities near the surface, resonance in cavities behind perforated surfaces or, in the case of light panels, friction caused by the flexing of the panel as it is set into vibration. In all cases sound energy is lost by transference to heat energy.

Thus, rough, porous or resilient materials will absorb more sound energy than those which are smooth, dense and rigid. Absorption varies considerably with frequency; in general porous materials absorb preferentially at middle and high frequencies while light panels absorb, by panel resonance, mainly at low frequencies.

Pores or cavities can of course be introduced into a material artificially—the familiar perforated fibreboard tiles being an example. Alternatively, an otherwise hard reflective material such as plywood can be perforated and mounted over an air cavity so that resonance takes place behind the panel. Such air spaces are often filled with a fibrous material, such as fibreglass, to increase still further the absorption of sound energy.

Measurement of Sound Absorption

Sound absorption can be expressed quite simply in terms of the percentage of energy absorbed compared with that reflected. Alternatively, it can be expressed as a coefficient, that is, total absorption as 1.00, 70% absorption as 0.70, 7% absorption as 0.07, etc. A list of absorption coefficients for common building materials and special sound-absorbing treatments is given on pp. 198 to 202.

However, for the purpose of the calculations described in sections 3 and 4, we need a *unit* of absorption by which the total effective absorption of a room interior can be expressed.

The unit of absorption still employed is that postulated by Professor Sabine towards the end of the 19th century. This he called the 'open-window-unit', that is, the absorption due to one square foot of open window—a 'surface' having 100% absorption. A square foot of surface having 70% absorption would therefore absorb 0.70 'open-window-units'.

In honour of his name the unit of absorption was subsequently called the 'sabin', or, more precisely, the 'square foot sabin' where surface measurements are imperial, and the 'square metre sabin' where measurement is metric. The 'sabin' can, however, be more generally defined as the absorption due to unit area of a totally absorbent surface. The relationship between percentage absorption, coefficient of absorption and square metre sabins is illustrated in figure 1.26.

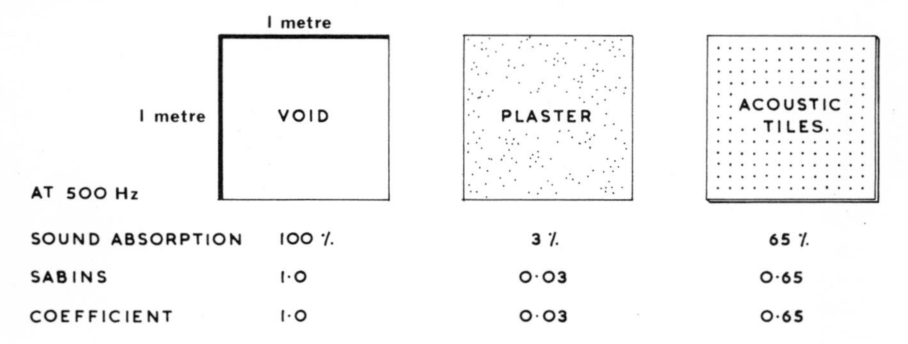

Figure 1.26

By using the sabin unit of absorption we can express the total absorption of a room by measuring the areas of all surfaces and multiplying each area by its appropriate coefficient of absorption. To the total of these values must be added the absorption due to any furnishings or persons present in the room and the absorption of the air itself. A typical calculation is given on p. 194.

In the case of people and seats it is usual to employ an 'absorption factor' rather than a coefficient, and multiply the absorption factor by the number of people or seats in the room. An absorption factor is the absorption due to the person or seat *as a whole* and is expressed in sabins. For this reason it is important to differentiate between square foot sabins and square metre sabins since the former is about one-tenth of the latter. A list of absorption factors for people and seats is given on p. 202 in both units.

2 Subjective Aspects of Sound

In section 1 we examined the nature and behaviour of sound *objectively* with little reference to human response and interpretation. In applied acoustics an understanding of the subjective aspects of sound is equally important if we are to discover design criteria for what is acceptable in the control of noise and the provision of good room acoustics. This section will therefore describe the physiology of hearing, the effects of noise on human well-being and the requirements for the appreciation of speech and music.

How the Ear Works

Figure 2.1 shows a simplified section through the mechanism of hearing. It consists of three parts: external, middle and inner ear. Closely associated with this mechanism, but omitted from the drawing, are the semi-circular canals which maintain the sense of balance. By reference to the drawing, the process

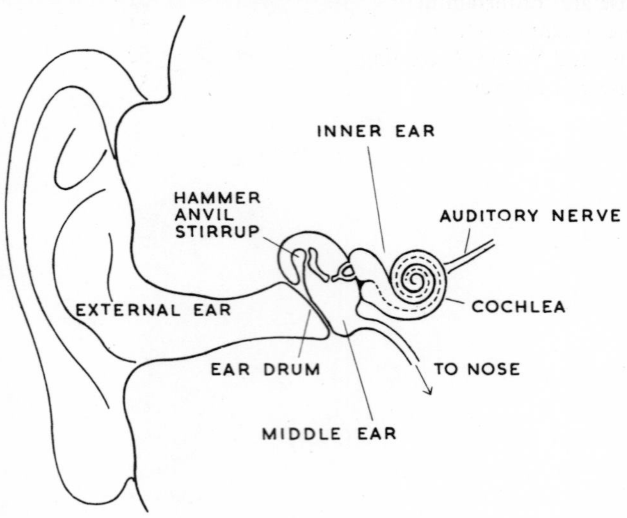

Figure 2.1

of hearing can be described in general terms as follows.

Sound pressure waves impinging on the ear drum (tympanic membrane) cause it to vibrate at the frequencies of the incident sound. These vibrations are amplified and transmitted across the middle ear by a mechanical linkage of small bones called, by reason of their shape, the hammer, anvil and stirrup. The cavity of the middle ear contains air maintained at atmospheric pressure by periodic connection to the mouth and nose during the act of swallowing. The base of the 'stirrup' seals one of two 'windows' into the inner ear or cochlea. Its vibrations are thus transmitted along the fluid contained within this spiral cavity.

Dividing the cavity of the cochlea along almost its entire length is a complex partition of bone and membrane to which some 25 000 nerve endings are attached. This partition is shown by a dotted line in the drawing. The nerve fibres from the receptor cells along the partition are gathered together to form the auditory nerve leading to the brain. Depending on the frequency pattern of the vibrations in the fluid of the cochlea, various parts of the dividing membrane are affected and the associated nerve endings excited. The vibrations are thus analysed and transmitted to the brain as codified messages, interpreted as sound of varying pitch and intensity.

Range of Response

The structure of the ear is such that it provides a natural defence against sounds of high intensity while being sensitive to movements of the receptor cells of microscopic dimensions. Nevertheless, explosive sounds of about 150 dB can rupture the ear drum or cause permanent damage to the nerve endings and associated parts of the cochlea.

Sounds of about 130 dB cause a tickling sensation, discomfort or even pain in the ear and this may be regarded as approximately the upper limit of tolerable hearing. Sound of much less intensity than this, if sustained over long periods for many years, can cause permanent deafness. The lower limits of hearing vary considerably with frequency but at 1000 Hz can be put at approximately 0 dB under ideal conditions and for pure tones. The ear's range of response in terms of intensity is summarised in figure 2.2.

In terms of frequency, the ear's range of response can be put at 20 to 20 000 Hz, again, under ideal conditions, for pure tones and in respect of young people with optimum hearing acuity. Age and other factors can reduce this range, mainly at the upper level.

Loudness and Frequency

The measurement of sound intensity level has been described in section 1. This is a purely objective measurement of sound energy and it is here necessary to relate this to subjective impressions of *loudness*. Sound levels expressed in decibels are not necessarily an indication of how loud sounds will seem to the human ear. This will depend on a number of factors, the most important of which is frequency, or frequency spectrum.

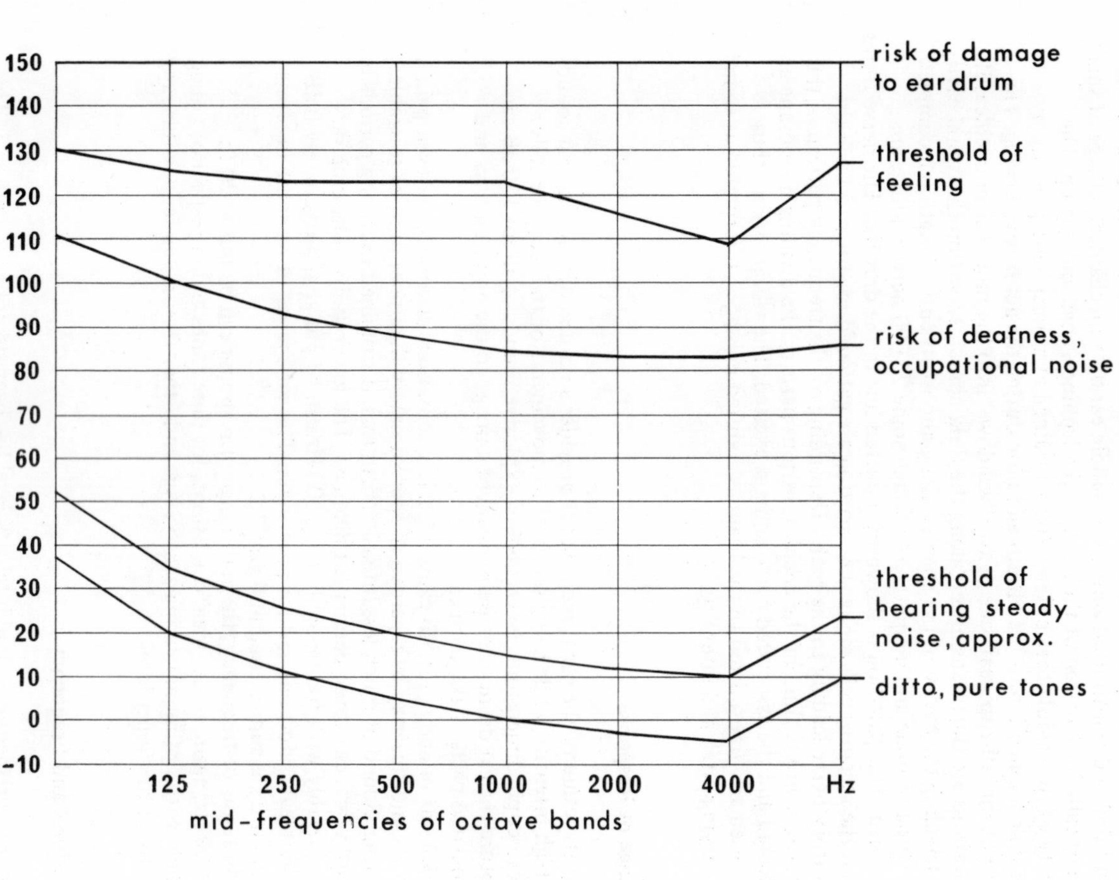
risk of damage to ear drum
threshold of feeling
risk of deafness, occupational noise
threshold of hearing steady noise, approx.
ditto, pure tones
150
140
130
120
110
100
90
80
70
60
50
40
30
20
10
0
-10
sound intensity level, dB
125
250
500
1000
2000
4000
Hz
mid-frequencies of octave bands

Figure 2.2

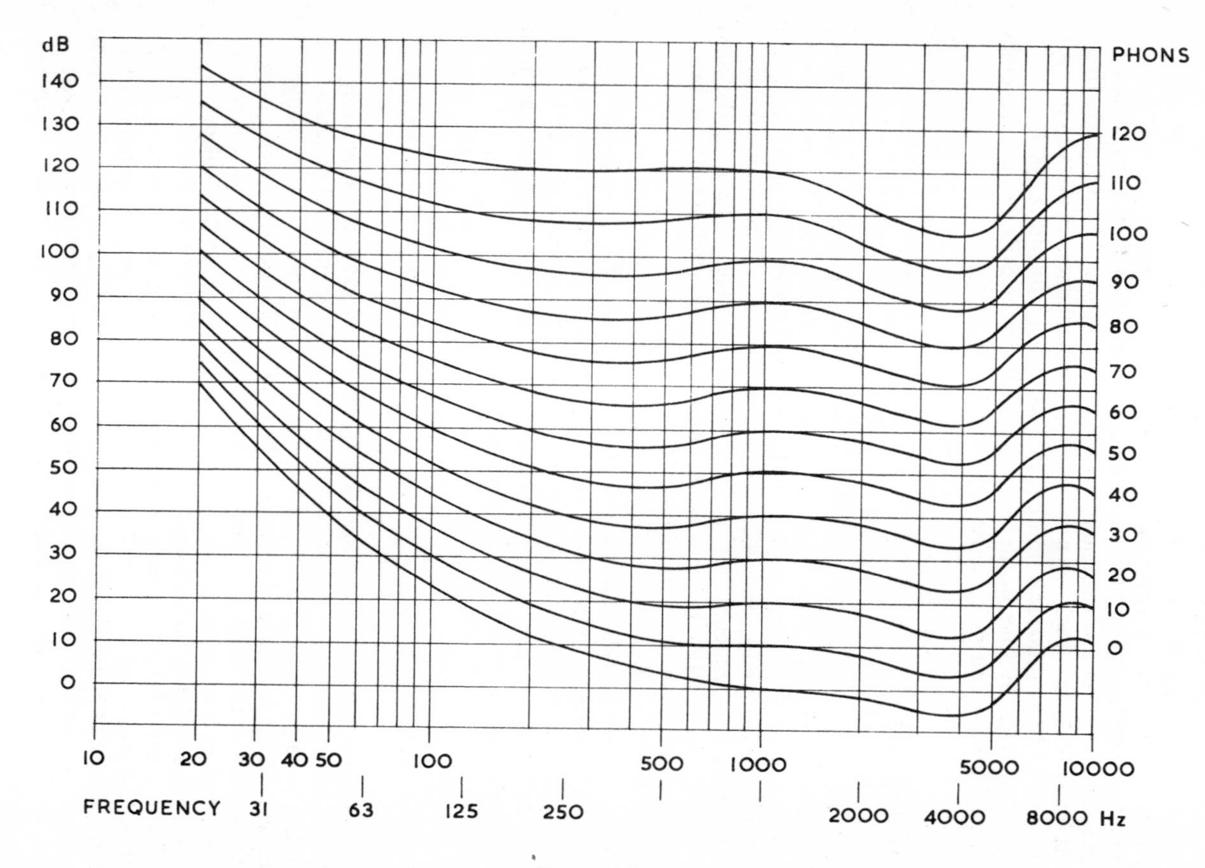
dB
140
130
120
110
100
90
80
70
60
50
40
30
20
10
0
PHONS
120
110
100
90
80
70
60
50
40
30
20
10
0
10
20
30
40
50
100
500
1000
5000
10000
FREQUENCY
31
63
125
250
2000
4000
8000 Hz

Figure 2.3

The sensitivity of the ear to *pure tones*—how loud we consider them to be—varies according to their frequency in a way which can be reliably ascertained. This relative assessment also varies with the intensity of the sounds being compared. Figure 2.3 explains this more clearly. The curves on the graph are contours of equal loudness determined by averaging the opinions of large numbers of people of good hearing. Thus a sound of 40 dB at 1000 Hz would be judged to be as loud as a sound of 50 dB at 100 Hz, or of 35 dB at 5000 Hz.

At higher intensity levels the assessment of equal loudness changes so that a sound of 100 dB at 1000 Hz is now equated with about 100 dB at 100 Hz and 90 dB at 5000 Hz. In other words, at the higher levels the ear's response becomes more uniform as frequency changes.

Scale of Phons

The above findings have made possible the formulation of a *scale of loudness*—the 'phons scale'—by which loudness can be given a numerical value. For pure tones, the scale of phons is shown on the right-hand side of the graph in figure 2.3. Thus if a pure tone is stated as being 40 phons, this is a measure of its loudness irrespective of its frequency. All pure tone sounds of 40 phons lie along the curve marked 40 on the graph, but their sound intensity levels, expressed in decibels, will vary according to their frequency.

The loudness of sounds of multiple frequency, such as noise, can also be stated in phons by a calculation based on the frequency spectrum. The calculation is complicated but the principle can be simply stated. If, in the case of a noise, sound of high intensity occurs in a band of frequencies to which the ear is particularly sensitive, this part of the noise will contribute more in loudness than if it occurred in an area in which the ear is less sensitive.

The dB.A Scale

An alternative means of measuring the subjective loudness of noise, and one which is widely used in practical work, is by means of sound level meters whose readings are adjusted to allow for the varying sensitivity of the ear to changes in frequency. Such meters have what are called 'weighting circuits' which can be switched in so that the meter's response to low frequency energy is reduced, to correspond with the reduced response of the ear, giving a reading which correlates very closely to loudness.

Some sound level meters have three weighting circuits, identified by the letters A, B and C, which can be switched in to suit the over-all intensity of the sound being measured. If these circuits are not switched in, the meter will simply indicate the sound pressure level in decibels (dB). When a weighting circuit is used, however, the reading is described as dB.A, dB.B, or dB.C and is a measure of loudness.

In the measurement of road traffic noise it has been found sufficiently accurate to rely on the 'A' weighting for noise control calculations so that, consequently,

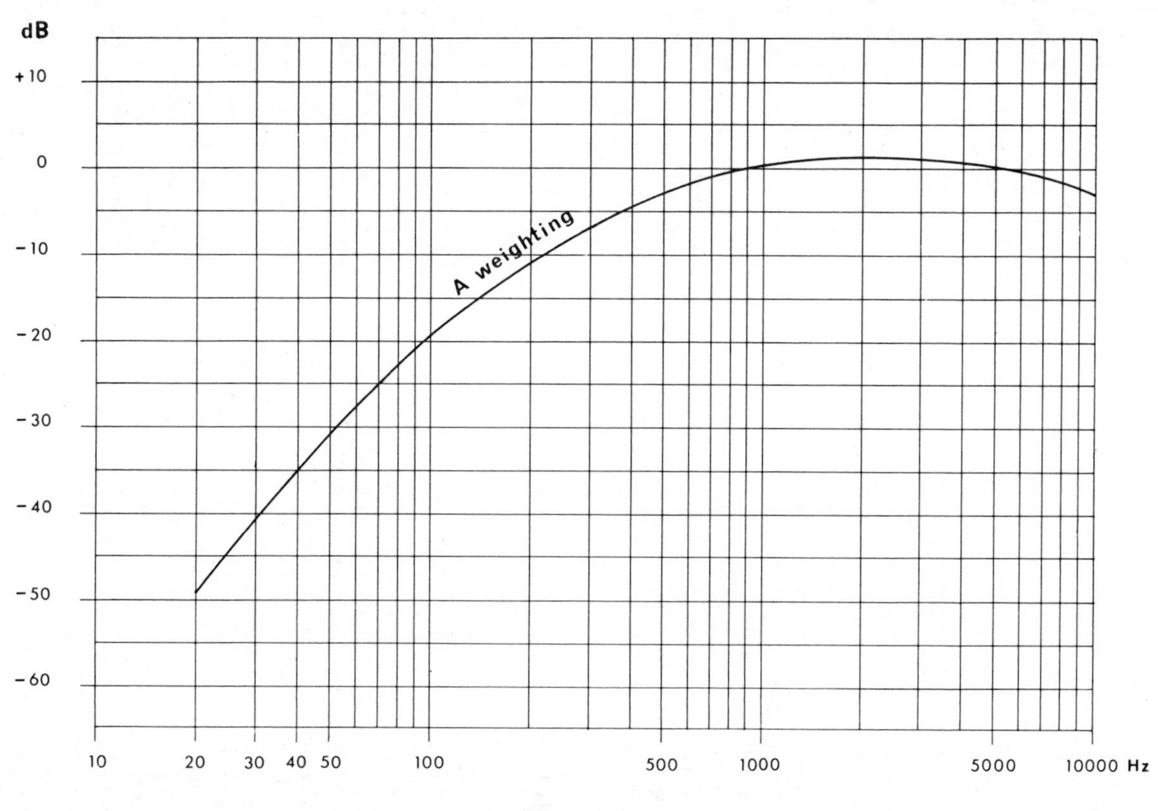
dB
+10
0
-10
-20
-30
-40
-50
-60
A weighting
10
20
30
40
50
100
500
1000
5000
10000 Hz

Figure 2.4

many meters provide only a dB.A scale of readings. Figure 2.4 gives the reductions in response of such a meter, relative to frequency. If this graph is compared with figure 2.3 it will be seen that the curve is roughly similar to a phons curve.

Threshold of Audibility

The threshold of hearing for pure tones is 0 phons but it has been seen from the graph in figure 2.3 that this corresponds to 0 dB only at certain frequencies. At 100 Hz, for example, a pure tone has to be raised to about 24 dB before it becomes audible. In the case of meaningless, steady noise the threshold of hearing is generally higher, as indicated approximately in figure 2.2.

In the general application of these two forms of measurement (decibels and phons), it must be understood that the thresholds of hearing stated above only apply under conditions in which all other sounds have been entirely excluded. This condition never obtains in practice and the lower limits of audibility of a given sound will be determined by the 'masking effect' of other sounds occurring in a natural environment.

This background or 'ambient' noise level varies considerably, from the effect of a light breeze passing across the ear on a silent mountainside to the hubbub of a works canteen. In practice it is therefore never necessary to reduce an intruding noise to the lowest limits of hearing to prevent it from being heard.

Masking Effect

The ability to hear a given sound will thus depend to a considerable extent on the prevailing climate of background noise. Depending on circumstances, the term 'background noise' may refer either to the noise generated around a listener, such as general room noise, or to noise entering the room from outside, such as road traffic noise. For example, general room noise will mask traffic noise so that it is less noticeable, or the traffic noise may mask, and make more difficult to hear, a conversation on the telephone.

It often happens that attempts to reduce internal room noise by acoustic treatment or the installation of quieter equipment results in external noise becoming more evident. The room noise had previously masked the external noise.

The masking of one sound by another is a complicated phenomenon. It depends not only on the relative intensities and frequency structures of the two sounds, but also on the mental attitude of the listener. However, when the dominant frequencies of two sounds approximately coincide the greatest masking effect occurs, and lower frequency tones have more masking effect on high frequency tones than vice versa. But also the degree of attention given by a listener to a certain sound will affect the degree of masking caused by another sound.

Background and Intrusive Noise

Although, as explained above, external noise and internal noise can both be termed 'background noise' in certain situations, it is useful to distinguish between them. Within a building the term 'background noise' usually refers to general room noise inevitably produced by the occupational use of the room. 'Intrusive noise' will then mean external noise, or noise from another part of the building, which penetrates the structural defences of the room. In practice, the considerations outlined above are taken into account in three ways

(1) Schedules of 'acceptable intrusive noise levels' have been developed as an approximate guide in design. These suggest reasonable criteria for various situations. For example, it may be said that intrusive noise should be reduced by structural insulation to 35 dB.A in the case of a hospital ward if patients are to be undisturbed by it, whereas an intrusive noise of 50 dB.A would generally pass unnoticed by the occupants of a busy restaurant. A schedule of such criteria is given on p. 112.
(2) An alternative, but approximate approach, is to reduce intrusive noise by structural insulation so that it is about 5 dB.A below the known or expected background noise level.
(3) A more accurate method of assessment has been developed where the frequency spectrum of the intrusive noise is known. In particular, this method has been applied in problems of speech interference to establish the degree of interference caused by noise according to its frequency characteristics. Examples of this form of calculation are given in section 3.

Effect of Noise on Health

There is as yet no reliable evidence that noise directly affects health, unless it is of such intensity as to cause physical damage to the mechanism of hearing. Apart from this, the effect on health is probably always indirect, for example, by causing loss of sleep or frustration in the carrying out of work. If this broader meaning is given to the matter, it can be said that the effect of noise on health may be serious.

During sleep muscular relaxation is almost complete, the heart rate decreases and blood pressure is lowered, the rate and depth of respiration is reduced and the nervous system is less active. Sleep is an essential period of physical and mental restoration and, if reduced in duration or depth over an extended period, physical and mental health suffers. This seems to be particularly the case with young children, whose physical and mental development can be retarded. The general effects of loss of sleep are diminished appetite, muscular weariness, lack of ability to concentrate or think rationally, and general irritability.

Depth of sleep is as important as duration, so that even noise insufficient to awaken the sleeper can be harmful over a long period. Moreover the time of day during which noise may disturb sleep is not confined to the hours of darkness; young children, night workers and the sick require sleep at other times so that, even where road or air traffic is reduced at night, the effect of the daytime noise cannot be disregarded. Nor is the effect on sleep directly related to the duration of the noise. One aeroplane passing over may cause considerable loss of sleep to a whole community and the babies awakened may reduce parents' sleep which cannot easily be made up.

A less obvious effect of noise on health is due to frustrations caused by interference with mental concentration. This applies equally to the business executive and the student preparing for examinations. Frustration is known to cause nervous disorders which react on physical health. Perhaps one other effect on health is worth mentioning, even if it cannot be measured. The increase of road and air traffic is making it increasingly necessary for people to keep windows closed during long periods of work and sleep. Most buildings provide no alternative means of ventilation. If medical opinion is correct in giving importance to fresh air, the long-term effect on health may be significant.

Noise and Deafness

Although the architect is seldom involved in the prevention of damage to hearing caused by sounds of high intensity, some reference to this aspect is necessary because, at those levels of noise where risk is marginal, acoustic treatment may be worth while. Otherwise the wearing of ear-muffs is essential.

Noises capable of causing permanent damage to the mechanism of hearing are typically those caused by explosions in quarrying, the testing of jet aircraft and certain industrial processes in which noise is amplified by panel vibration. Examples of the latter are boiler riveting and the grinding of large propeller blades. Apart from explosions that are capable of causing immediate damage, what is called 'occupational deafness' can be caused by noise of the order of 90 dB.A if suffered over an extended period.

Short exposure to loud noise causes temporary deafness, over part of the frequency range, lasting anything from a few seconds to a few days. Prolonged exposure during a working life produces permanent and, as yet, incurable deafness. Hearing loss is at first in the high frequency range and is unnoticed. It then extends over lower frequencies until the important speech frequencies are seriously affected and the understanding of conversation becomes difficult or impossible.

The risk of such hearing loss varies from individual to individual and according to the frequency spectrum of the noise and whether it is intermittent or continuous. Various authorities have suggested limits of noise level above which the ear should be protected, and those given in figure 2.2 are due to Professor Burns and Dr Littler. It should be noted that these levels are not remarkably high. The criterion applies to broad band noise sustained for eight hours a day, five days a week. If the noise in any octave reaches the level shown on the graph, ear plugs or preferably ear muffs should be worn.* Ear plugs vary greatly in efficiency and

*A full treatment of this subject is given in *Code of Practice for Reducing the Exposure of Employed Persons to Noise* (Department of Employment, H.M.S.O., 1972).

are, at best, less effective than fluid-seal ear muffs, which can reduce noise by 10 to 40 dB according to frequency. Impulsive noise, or noise containing intense pure tones, are thought to present a greater risk.

Such noise will not only affect the people immediately concerned but also those working at a distance. Here the architect may be required to suggest defence measures in the form of screens, dividing walls or sound-absorbing treatment. These possibilities are discussed in section 3.

Speech Intelligibility

The understanding of speech is dependent on two qualities: power and clarity. It is obvious that at the point of reception all, or nearly all, speech sounds should have sufficient loudness for oral interpretation. This requirement presents a major problem in the design of large auditoria, especially when it is borne in mind that some of the discrete sounds in speech have very little power at the outset. It is probably true to say that, even in a medium-size auditorium a listener at the back of the room is dependent on 'guessing from context' for some of the words spoken. Inevitably there will be a loss of power due to distance but, as will be discussed in section 4, further losses can be minimised by good design.

In many rooms of modest size, however, the understanding of speech is poor, not because of lack of power, but because of lack of clarity. We have already mentioned the masking effect of ambient noise and intrusive noise. This in effect causes a loss of clarity since the actual power of the speech is unaltered. There is, however, an additional phenomenon which occurs when speech is heard in an enclosed space, and which can reduce clarity, namely, delayed reflections.

If we refer to figure 4.20, we will see that each discrete speech sound is represented by a peak on the graph followed by a slope representing the decay of reverberation for each sound. The angle of slope represents the rate of decay, or the time taken for each sound to lose power. The longer the reverberation lasts, the more speech intelligibility will be confused by the filling in of the gaps between each sound and the masking of subsequent sounds.

On average, the duration of a syllable in speech is 1/5 s and the gaps between words about 1/3 s. The more powerful components of reverberation can therefore effectively mask consecutive sounds and fill in the gaps between words. This loss of articulation or 'blurring' is illustrated in figure 4.20 and, for this reason, rooms for speech should be designed so that reverberation decays rapidly and the more powerful primary reflections follow quickly upon the direct sound.

The above aspects of power and clarity—how they can be maximised in the design of auditoria—will be discussed in some detail in section 4.

Noise and Speech Interference

The brain has remarkable powers of discrimination in the oral as well as the visual sense. It can accept 'wanted' sounds and reject others, even if the latter are comparable in loudness. At certain levels of intrusive noise speech can, however, become unintelligible and, even below these levels, the strain of listening may become insupportable. The criterion is not whether speech can be heard but whether it can be understood without undue strain.

The general effect on speech of ambient or intrusive noise is also illustrated by the graph in figure 4.20 by the horizontal line marked 'ambient noise'. Parts of speech are submerged (masked) by the noise and ability to understand will depend on how much remains above the horizontal line and to some extent on what the brain can decipher of that which lies below. It will be aided by our ability to fill in the meaning of words from a knowledge of their context.

It appears also that the extent to which a noise interferes with the understanding of speech depends not only on its loudness (as may be expressed in phons or dB.A) but also on the particular pattern or spectrum of frequencies of the noise. In the assessment of 'speech interference' the noise is analysed in octave bands and then this is related to the level of the voice and the distance between speaker and listener. This method is described in section 3.

Other factors, as would be expected, have their influence and for some of these a reasonable allowance can be made, for example, whether the noise is continuous or impulsive, whether it is meaningless or not, and whether it is pleasant or irritating.

Effect of Noise on Efficiency

The effect of noise on working efficiency depends on the nature of the work being performed, even to the extent that some tasks may be better performed in an environment of moderate noise than in a quiet one.

Experiments to assess changes in working efficiency due to changes in acoustic conditions are very difficult to confirm under actual working conditions because of the impossibility of entirely excluding other factors which may affect the result. It has been discovered, for example, that sound-absorbent treatment of a workshop has on occasion improved efficiency, not because noise was reduced, but because the morale of the workers was raised by the interest taken in their welfare.

Because of this, there is some disagreement between the results of the many experiments carried out along these lines in industry. There is, however, general agreement that the hopes entertained for a marked and *continuing* increase in efficiency as a result of noise reduction have been over-optimistic. For most factory operations it seems that not until noise levels around 85 dB.A are reached do material changes in over-all output occur, and that reductions in efficiency are then not more than about 5% and generally much less.

This is not to suggest that factory noise should not be reduced since, as we have seen, it is at noise levels around 85 dB.A that the risk of occupational deafness occurs. However, if we consider efficiency in isolation, controlled laboratory experiments and work in the field indicate that

(1) Efficiency in the performance of monotonous, automatic work may be improved in the presence of a moderate amount of noise (or music) owing to an effect of 'mental arousal'.

(2) Noise of about 85 dB.A or more may increase the number of errors made in skilled manipulative work, owing to an effect of psychological agitation.

(3) Work requiring mental concentration, creative thought, response to signals and the taking of decisions is affected by noise well below these levels. It is thought that noise competes with other sensory processes, requiring the subject to turn his attention to it periodically, during which time errors occur. In the case of low-level noise, that is, below 60 dB.A, efficiency is only affected by meaningful (information-providing) sources, for example, a background conversation. Random noises such as typewriters or ventilation system noise do not have any material effect on efficiency in the case of clerical work.

(4) The degree of disturbance depends on whether the noise is meaningless or not, the subject being disturbed by any subconscious effort to interpret meaning.

(5) All kinds of work are adversely affected at noise levels of 90 dB.A and above.

Noise and Accidents

Because noise increases the possibility of errors in both skilled manual work and some kinds of mental work, the chances of accidents occurring have also to be considered. The effect of noise on over-all output may be negligible in some types of work, or it may be significant in processes where a mistake holds up a sequence of operations. This may be said to be chiefly the concern of management, whereas mistakes which lead to accidents directly concern the operatives, if not others who depend on them for their safety.

If, as seems to be the case, response to visual signals is less reliable under noisy conditions, then those operations on which the safety of people depend should be carried out in a quiet environment. The operation of some types of machine requires the operative to be able to hear the changes of sound produced in order to effectively control it. It may therefore be necessary to screen him from the noise of other machines. Probably the most likely cause of accidents is the misinterpretation of verbal instructions, direct or by telephone, and here the assessment of speech interference, already referred to, is important.

It is known that in a noisy environment, some people are much more prone to make mistakes than others and that some people are practically unaffected. Where the results of mistakes may lead to accidents, it will be the reactions of sensitive people that must be taken into account, rather than any 'average' response.

Noise and Annoyance

The extent to which noise irritates people (and the reason it does so) is the most intangible aspect of the subject of noise control. The physiological and psychological factors are so complex and vary so much from individual to individual that any theoretical attempt to anticipate people's reactions to noise seems impossible.

Most information regarding this form of noise disturbance is based on social surveys and the complaints received by local authorities. Some attempt has been made to assess the likelihood of future complaints on this basis (see p. 133). But, since complaints regarding annoyance are inextricably mixed with complaints of interference with sleep, reading or conversation, the 'annoyance ingredient' alone is difficult to assess. This aspect is, however, no less important because it cannot be measured or fully explained.

It is therefore worth while to enumerate some of the factors which appear to determine the degree of annoyance caused.

(1) Worried, sick and psychologically disturbed people seem to be most affected.

(2) There are considerable differences of susceptibility to noise in otherwise 'normal' people, probably connected with emotional character, stamina and general outlook.

(3) In certain circumstances, some people find noise exciting and emotionally satisfying, others hardly at all.

(4) In comparable circumstances, young people are less likely to be irritated than older people.

(5) The level of noise to which people have become accustomed will influence their attitude towards it.

(6) People are more likely to complain of a new noise than of one they have heard before.

(7) If people think of a noise as being unavoidable, they may be less irritated than if they consider it unnecessary.

(8) Other associative ideas seem to play a part, for example: personal dislike of a neighbour who happens to make the noise, associations of fear, or unsatisfied curiousity.

(9) The degree of annoyance caused is not directly related to its intensity. Its frequency spectrum, periodicity, and the information contained in it may be more important. Generally, high-frequency, harsh or unmusical sounds, sounds of fluctuating pitch or unpredictable rhythm cause most annoyance.

It is with the above in mind that the criteria employed for practical purposes of noise control must be judged. Such criteria are suggested in section 3. Most of them are based on a statistical analysis of public opinion and do not take account of minorities who for one reason or other may react 'abnormally'.

Effect of Noise on Learning

The effect of noise on education is obvious where interference with the understanding of verbal instruction results. It perhaps need only be said that noise does not have to be continuous for its effect to be serious. Many processes of teaching depend upon the ability of the teacher to take his pupils through a sequential train of thought and occasional distractions (such as aircraft) may have an effect out of all proportion to their number.

The effect on the mental concentration involved in personal study has already been mentioned, but it should be added that experiments in industry have shown that learning a new job, not necessarily requiring great intellectual effort, may take longer in a noisy environment than in quiet conditions.

Noise and Entertainment

That extraneous noise can reduce the enjoyment of music, drama, television and other entertainments is evident in itself but the *extent* to which noise must be excluded is not simple to determine. There are three factors to be considered

(1) the range of loudness of the sound listened to,
(2) the level of unavoidable room noise to be expected and
(3) the character of the intrusive noise.

Criteria for design are suggested in section 3 but in general terms it may be said that where considerable variation occurs in the loudness of the sound to be appreciated, the level of intrusive noise requires greater reduction. Thus in a theatre, where the actor's voice may fall to very low levels for listeners in rear seats, insulation from external noise is more critical than in, for example, the cinema. Likewise, the concert hall requires very quiet conditions for the enjoyment of pianissimo parts of the score. Where ambient room noise is unavoidably high, as in a dance hall, the reduction of intrusive noise need not be so much. And, again, steady 'broad band' noise requires less reduction than impulsive, erratic sounds of fluctuating frequency spectra. The criteria given later should be viewed in this light.

Musical Appreciation

Where standards are high, the composition of music and its interpretation by conductor and players involve subtleties of communication of a very sophisticated nature. Moreover we are dealing with sounds which are extremely complex. Each note of music in an orchestra is made up of a fundamental tone and a number of harmonics, often preceded by a 'transient' (an initial sound of different character from the sustained note). These notes, in the case of some instruments, are combined into chords and we may have as many as eighty instruments of various kinds playing in unison.

For the full appreciation of music we therefore require a proper hearing of this complex sound with all its subtleties of arrangement and interpretation. When, however, such music is played in an enclosed space to a large audience both the enclosure and its occupants change the character of the sound heard by a member of the audience.

A well-known example of the effect of an enclosure on the sound heard is of course the possibility of an echo, heard, incidentally, by some members of the audience and not others. This is only an extreme example of a delayed reflection. In any room, however, delayed reflections set up reverberation which, if excessive, can detract from our enjoyment of music. The presence of a large audience can, in itself, modify the power and distort the frequency balance of the music produced at the platform, especially for listeners at the rear of the auditorium. Even the arrangement of the audience in relation to the platform can modify the reception of the music as intended by the conductor since people seated well to the side of a large orchestra will not hear the balance of instruments quite in the same way as those seated behind the conductor.

The general aim therefore in designing a room for music is to provide all members of the audience with, as near as possible, the same quality of sound as is heard by the conductor in rehearsing his orchestra, and an acoustic environment appropriate to the type of music being played. The design implications of such an objective are considered in section 4.

Noise and Sound Recording

Whereas most people will not object to occasional and moderate noise in the environment in which they listen to a gramophone, the radio or television, they will certainly complain if extraneous noises are emitted by the disc or broadcast over the air. For this and other reasons the criteria for noise exclusion in studios are higher than in any other situation, except acoustic laboratories.

Noise and Privacy

Architects, in the planning of buildings, need no reminder of the need for visual privacy and the thinnest of opaque materials will provide it. The provision of acoustic privacy seems, however, to be sometimes confused with the requirements for visual privacy and the barriers to hearing may be quite inadequate. It is of course unfortunate that, whereas a small sample of material can immediately be seen to be opaque to light, the assessment of a material's acoustic opacity is far more complicated. It involves not only the material in question but also the whole construction in which it occurs. It requires a knowledge of sound levels, insulation values and criteria, sometimes at six frequencies.

It is already evident that the prefabrication of light panel materials is reducing standards of acoustic privacy in the home, the office and the consulting room. This applies not only to the design of partitions between rooms but also in the design of external claddings which act as alternative paths of sound transmission.

3 Noise Control

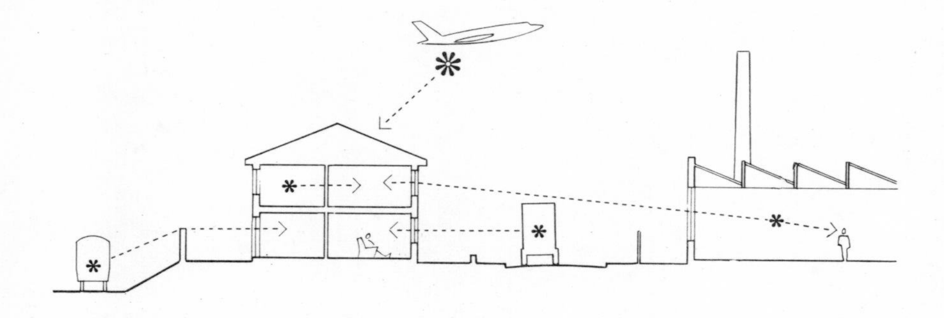

Design Principles

In this section the general principles involved in reducing noise, or defending people from it, will be explained and related in broad terms to design. It must, however, be stressed that a knowledge of principles is only the first step in design. Without also some means of quantitatively assessing the amounts by which noise can be reduced and, without criteria for the levels of noise which are acceptable, a knowledge of principles is of little value.

Such quantitative assessments may be, as generally in the past, a matter of personal experience, or they may be by calculation. The kinds and intensities of noise are, however, changing rapidly, as also is the design and construction of buildings. Judgements based on personal experience alone have become increasingly inadequate in the anticipation of results. Some means of calculation are necessary. For convenience of reference, however, principles will be separated from methods of calculation and the latter will, in the main, be dealt with in the latter part of this section.

Reduction of noise at the source

The architect will sometimes have the opportunity of preventing noise 'before it starts'. To the extent that he is able to choose equipment or advise his client regarding the installation of machinery, he can exert some control over the introduction of unnecessary noise into his building. There is considerable variation in the noise produced by items of equipment performing similar functions. To mention only a few: there are quiet and noisy versions of ventilation fans, flushing cisterns, cleaning equipment and some types of industrial machinery.

The production of noise can usefully be considered under three headings

(1) sustained vibration of solid bodies
(2) impact vibration
(3) aerodynamic noise.

Sustained vibration of solid bodies

Examples of sustained vibration of solid bodies are the electro-physical activation of a loudspeaker cone, the rotary or reciprocating motion of a machine and the frictional vibrations induced by a band-saw. The general characteristics of such sounds are that, as long as the speed of movement and the force exerted remains constant, the frequency spectrum and level of noise also remains constant. A reduction in the speed of movement will in general lower the pitch and, because the ear is less sensitive to lower frequencies, reduce the loudness of the sound. In all cases the production of audible sound will depend on the presence of surfaces large enough to propagate air-borne sound waves.

It follows then, that in the choice of equipment, that which entails a slower speed of movement is likely to be quieter. Where panels which act as amplifying diaphragms are necessary, they should be 'damped' by the application of viscous material or be heavy enough to reduce vibration. In some cases such panels can be omitted, as in the substitution of wire guards for sheet metal around the moving parts of lathes. Frictional noise can often be reduced by improved design and lubrication. Self-lubricating and vibration-absorbing nylon gears can sometimes be substituted for steel gears. Rotary movement in machines is usually quieter than reciprocating action.

Impact noise

The vibrations caused by impacts, such as footfalls or slamming doors, are transient; the vibrations quickly decay. Nevertheless, they usually contain considerable amounts of energy and can travel long distances through solid structures with little attenuation. Such vibrations are usually communicated directly to a resonant panel, such as a floor or partition, to produce audible sound. In addition they travel through structures to set other panels into vibration, sometimes at a considerable distance from their point of origin.

Impact vibrations are most effectively checked at the point of impact. Soft floor finishes and rubber stops on door frames are more effective than sound-absorbing material elsewhere. When this is not possible, as for example in the case of a gymnasium floor, the next best precaution is to prevent, as far as possible, the impact vibrations from reaching the structure. This method has given rise to the concept of the 'floating floor', in which the floor finish is acoustically isolated from the structural floor *and the walls* by a resilient material. The system is illustrated in figure 3.1. What is called 'edge isolation' is as important as the horizontal separation. At no point must the 'gap' be bridged by such elements as skirtings.

The choice of resilient material is important. It is required to be elastic, that is, not subject to permanent compression, and viscous, that is, offering resistance to molecular deformation. The material should be employed 'under load' to damp out vibrations, so that the floor finish itself should have sufficient weight. Finally, the material must maintain its resilient characteristics with age.

Apart from the surface treatment of the floor, the weight of the structural floor will also influence impact sound transmission to the room below. Hollow structural floors do not, as is sometimes thought, provide superior insulation against impact (or air-borne) noise, because they are usually lighter than their solid counterparts. While a soft floor finish provides the best reduction of impact vibration, a combination of soft finish, floating floor and heavy structural floor will give the best over-all result.

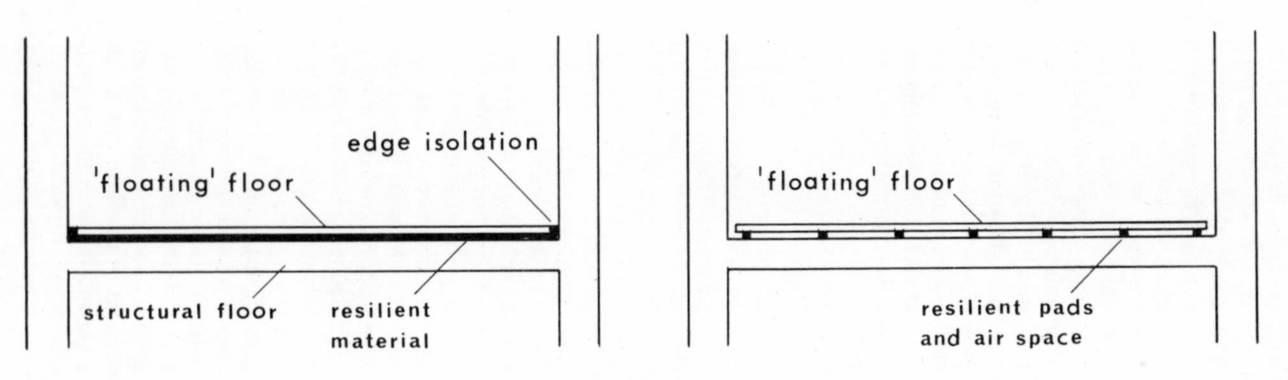

edge isolation
'floating' floor
structural floor
resilient material
'floating' floor
resilient pads and air space

Figure 3.1

Aerodynamic noise

Aerodynamic noise is produced in a number of ways all of which involve either turbulence or the sudden expansion of gases, sometimes a combination of both. A familiar example of noise due to air turbulence is that produced by a propeller or fan blade in which pressure waves are set up of varying frequency depending on the speed of the blade. In general, low-speed fans are 'quiet' because the ear is less sensitive to the lower tones produced and because the pressure waves are less intense.

A similar phenomenon of 'edge turbulence' is produced when air is blown across the edge of a solid body. Thus rapid air movements or escaping gases will cause noise if associated with the partial obstruction of solid elements. The jet engine is an example of a source of noise due to the formation of vortices, resulting in noise levels of around 130 dB measured at 40 m.

The second category of aerodynamic noise is the sudden expansion of gas, such as occurs in an explosion, which sets up a pressure wave of, normally, very high intensity. For example, the sound level of the report from a 7.5 cm cannon is about 140 dB at 5 m.

Repeated explosions, such as occur in an internal combustion engine, create a sustained noise by way of the exhaust system, in addition to setting up vibrations in the associated parts of the engine. Add to this the air turbulence of the exhaust outlets, the vibration of body panels together with tyre friction and it will be seen that the motor vehicle embodies most if not all the possible causes of noise.

In conclusion, it may be said that any additional cost involved in the reduction of noise at the source is likely to be far less than the expense of sound insulation once the noise is produced.

Reduction of noise near the source

Noise or sonic vibrations, once produced, are best checked as close to their point of origin as possible. In a few cases this may be done by the complete enclosure of the noise-producing element, together with its 'acoustic isolation' from any supporting structure. Thus a noisy machine may, if the need for access permits, be enclosed and mounted so that all or sufficient of the noise and vibration is prevented from reaching the surrounding air and structure. Figure 3.2 illustrates diagrammatically such an enclosure with the possible paths of sound transmission marked by numbered arrows.

Path 1 is transmission by way of the enclosing panels, which must be heavy enough to provide sufficient insulation. Arrow 2 indicates reverberation within the enclosure, which can be reduced by a sound-absorbent lining. Path 3 is the transmission of vibrations to the floor structure, which can be absorbed by a resilient mat or coil spring mountings. Path 4 is 'air-to-air transmission' through any openings (such as vents) in the enclosure. Such openings can be provided with acoustic baffles as shown.

In any sound insulation problem it is essential to locate all possible paths of sound transmission. If one path is overlooked this may negate the precautions taken with regard to the others.

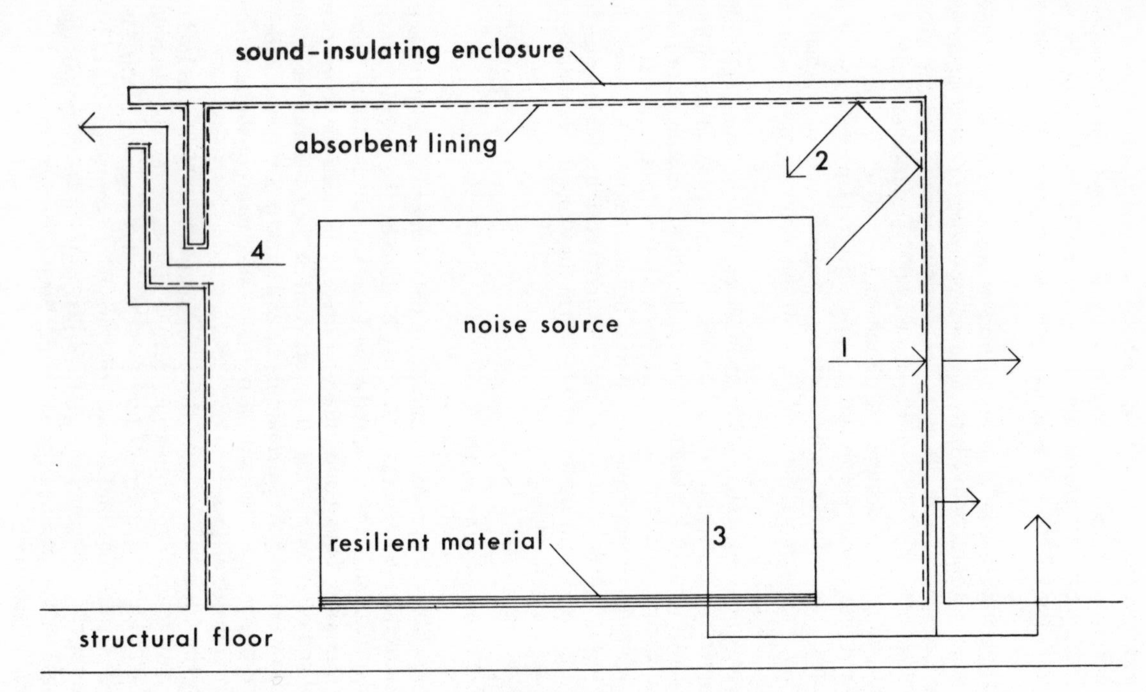
sound-insulating enclosure
absorbent lining
2
4
noise source
1
resilient material
3
structural floor

Figure 3.2

Reduction of internal noise by screening

To a limited extent, reduction of noise may be achieved by partial enclosure or screening. For example, hand-operated lathes may be acoustically separated from each other by screens so that *direct* sound from one machine is not heard by the operative at another. It may be possible in this way to meet the need of the operative to hear his own machine above the noise of others, besides providing better working conditions.

Screening will, however, have little effect if the need to reduce reflected and diffracted sound is overlooked. Figure 3.3 shows an arrangement of screens between lathes and the paths by which sound can travel from one bay to another.

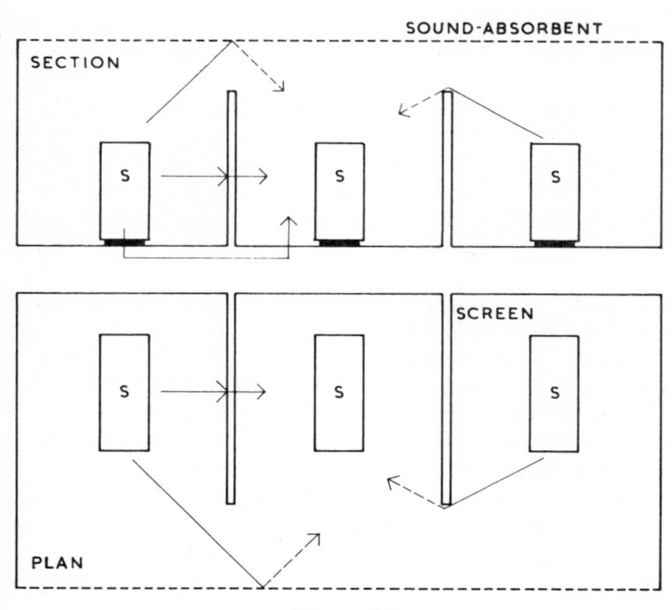

Figure 3.3

The general requirements are seen to be

(1) that the screens should have adequate sound-insulation properties
(2) that absorbent material should be used to reduce reflected sound
(3) that the screens should be as large as possible, to reduce diffracted sound and
(4) that transmission of vibrations through the floor should be damped out by resilient mountings.

If practical, other surfaces, including the screens, may be surfaced with absorbent material to further reduce the general noise level in the workshop. This aspect of noise reduction is discussed below.

Reduction of room noise

There are two reasons for reducing reverberant room noise: (1) for the comfort of the occupants and (2) to reduce the noise transmitted to adjoining rooms.

We have seen (p. 34) that noise produced in a room can build up to a level which is determined by the absorption characteristics of the room. This build-up may be reduced to negligible proportions if the room contains a sufficient amount of sound-absorbing material. Furthermore, any transient noise, such as the slamming of a door, will be absorbed quickly (as far as air-borne sound is concerned) and the room will be that much quieter. In addition, conversation will be made easier by the reduction of reverberation, which tends to reduce the clarity of speech.

It must, however, be made clear that no amount of acoustic treatment can reduce the noise in a room below the level it would be if heard in the open air—it can only prevent an unnecessary increase.

For the occupants of adjoining rooms the reverberant noise level in a room is also important. The degree of sound insulation required between two rooms will depend on the reverberant sound level in the 'source room'. It follows then that if the reverberant sound level can be reduced by acoustic treatment, less insulation will be required. In the calculation of insulation requirements reverberant sound level is the starting point in all cases where the source of sound is enclosed.

All rooms possess a certain degree of absorption, whatever materials are used, and most rooms are occupied by people who will effectively absorb sound. Whether the build-up of reverberant noise can be further reduced by acoustic treatment will depend on what *proportional* increase can be made in the sound absorption already present. The calculation is given on p. 130, but as an example: it is necessary to *double* the total sound absorption in a room (as expressed in sabins) to effect a 3 dB reduction in reverberant sound level.

If a room has mostly hard surfaces, no soft furnishing and few occupants, it is not usually difficult to make a fourfold increase in absorption, and obtain a reduction of 6 dB. If, however, a room is already very absorbent, such as a well-furnished living room, it may be impossible to make any material improvement. Before undertaking to reduce reverberant noise in a room it is therefore essential to calculate the amount of absorption already present in order to find out *by what proportion* it can be increased.

Application of sound-absorbing material

In the application of special sound-absorbing material a lot will depend on the use to which the room is put and the availability of surfaces for treatment. In rooms occupied by a number of people and with a certain amount of sound-absorbing furniture, noise produced at a low level (such as from typewriters) will to some extent be absorbed on the horizontal plane. It becomes therefore more important to absorb sound travelling upwards. The ceiling and upper part of wall surfaces are then more effective areas for treatment, as in figure 3.4.

Acoustic absorbents of high efficiency are obviously necessary if the cost of treatment is to be justified but, in addition, the coefficients of absorption of the

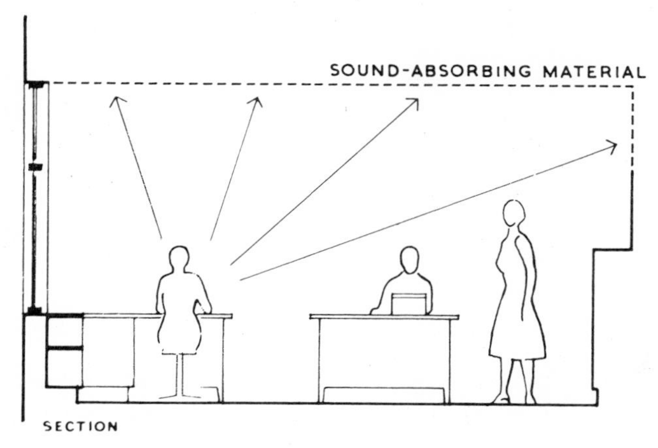

Figure 3.4

material at various frequencies must be examined. For example, if the noise in question is predominantly of low frequency, then the material employed must be effective in this region. Some proprietary materials have very low coefficients at around 125 Hz.

There is sometimes difficulty in finding space for a sufficient surface area of absorbent material, so it should be mentioned that acoustic material need not necessarily be applied 'on the flat'. Surface area, and therefore total absorption, can be increased by modelling the available surface or, if no surface at all is available, sound-absorbing units can be suspended from the ceiling. In cases where a laylight occupies the whole ceiling area, the sound-absorbing material may be

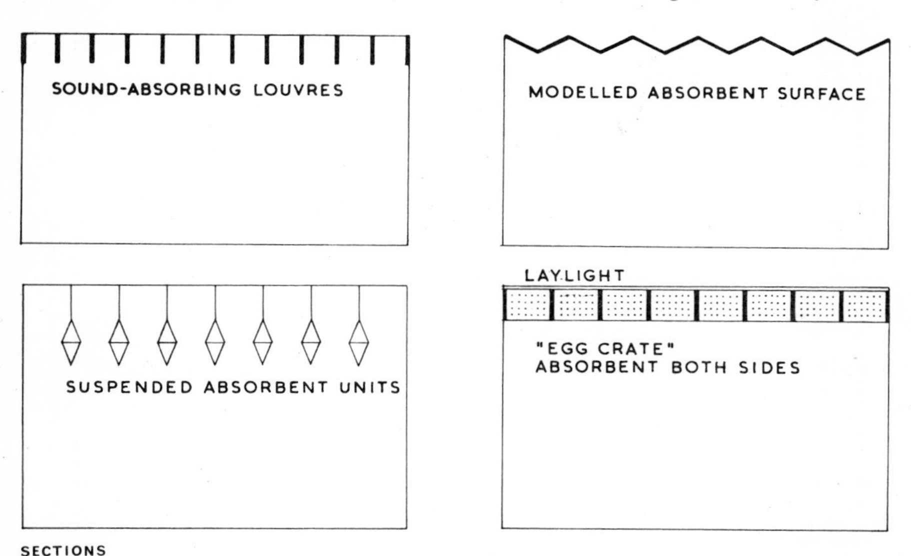

Figure 3.5

applied in the form of louvres (or 'egg-crate' arrangement) without materially obstructing the entry of light. Both sides of the louvres can be made absorbent. Some of these arrangements are shown in figure 3.5.

It is sometimes thought that by applying sound-absorbing material to a partition between two rooms the transmission of sound between the rooms will be reduced. Since the weight of the partition is not materially increased, this is not so. The presence of the material will reduce the reverberant sound level in the room and this is an advantage, as explained above, but there is no particular merit in applying the material to the dividing partition.

Reduction of noise by structural defence

Having considered the possibility of reducing noise at, or near the source, we may next consider the defence provided by building structures to both internal and external noise. Internal walls, partitions and floors can defend the occupants of a building from sound generated in other parts of the building, such as from machinery, equipment, music or speech. External walls, windows and roofs can defend the occupants of a building from noise generated outside, such as that from road, rail and air traffic or nearby factories.

In what follows we shall be referring only to defence against air-borne sound.

The defence provided may be termed the 'insulation value' of the structure, and can be expressed quantitatively as the difference in sound level between one room and another, or between the exterior and the interior of the building.

Figure 3.6 illustrates three situations. Example A may be taken as representing road traffic noise at a level of 75 dB at the external face of a building, reduced

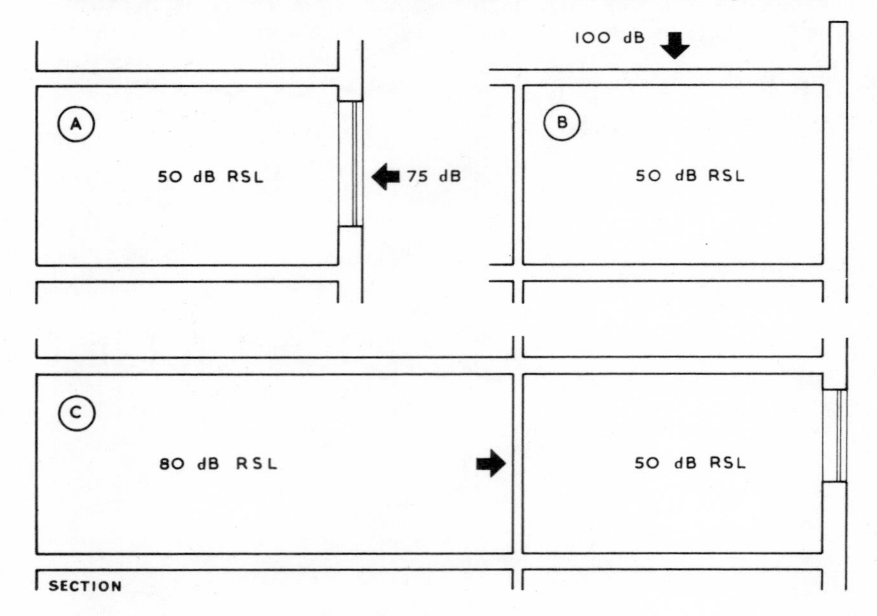

Figure 3.6

to an average 50 dB in the room shown (it will be higher near the window and lower away from it). The insulation value of the structure (not the external wall and window alone) may therefore be said to be 75 – 50 = 25 dB.

Similarly, in example B aircraft noise of 100 dB is reduced to 50 dB in the room, indicating a reduction by insulation of 50 dB for the roof and adjoining structure. In example C noise is being generated in one room (possibly by a machine) and reaches a reverberant sound level of 80 dB. If the average sound level in the adjoining room is 50 dB then the insulation value of the partition and related stucture is 30 dB.

In this method of defining insulation value no account is taken of any additional noise which may be produced in the defended room—for this purpose it is assumed that there is none. It should also be noted that the insulation value of a structure remains constant. Thus, if in any of the above examples the intrusive noise level is raised, the noise level in the receiving room will also rise by the same amount.

Direct and indirect sound transmission

If we consider any of the examples illustrated in figure 3.6, we may say that sound transmission is of two kinds

(1) direct transmission by way of the barrier between noise source and listening position and

(2) indirect or 'flanking' transmission along adjoining elements of the enclosing structure.

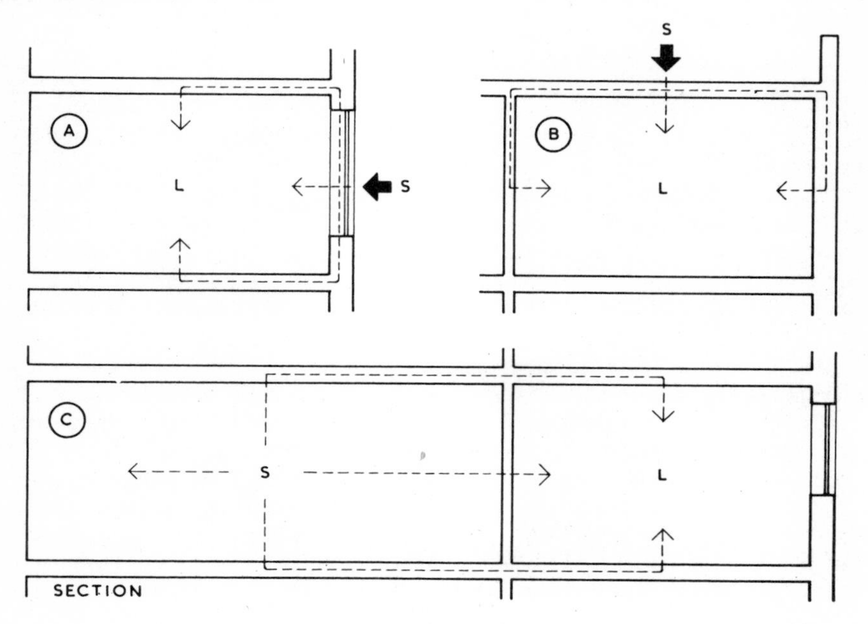

Figure 3.7

Figure 3.7 indicates by arrows some of the paths of indirect transmission in the three situations. The vibrations set up, for example, in the roof of case B will travel down the external wall and the internal partition, to radiate sound into the room. In fact all the bounding surfaces of the room will, theoretically, contribute to the resultant sound level in the room. How much each element will contribute depends on its mass, its position and the rigidity or otherwise of the interconnections between elements. Having in mind the complicated nature of most building structures—the changes in material, profile, weight and junctions—it will be appreciated that the sound insulation of a complete structure can only be *estimated* from a knowledge of the insulation values of its parts. It is not subject to exact calculation. If, however, a barrier between sound source and listening position has a known insulation value then, provided flanking elements are comparable in mass, that insulation value can be assumed to apply to the whole structure. It is in those cases where flanking elements are much lighter than the dividing element that effective insulation can be reduced by 'flanking transmission'.

Measurement of sound insulation values

A schedule of sound insulation values for various building elements is given on p. 111. It is necessary to understand exactly what is being measured and something about the methods of measurement.

Firstly, as explained in section 1, any panel acting as a barrier between source and listener will reduce the intensity of the sound by a constant *ratio*. Insulation values can therefore be expressed in decibels.

The sound insulation values (or sound reduction factors) of building elements can be measured in two ways

(1) under controlled laboratory conditions or
(2) as used in buildings.

Under laboratory conditions indirect (flanking) transmission is eliminated by building a sample panel into a more massive construction so that the inherent insulation of the panel can be discovered. The panel is built into a wall between the 'source room' and the 'receiving room', both being reverberant chambers of known acoustic character. The difference in sound level between one chamber and the other gives the sound reduction factor in decibels. By analysis of the sound on each side of the panel insulation values can be ascertained for a range of frequency bands.

Where measurements are taken in buildings, however, flanking transmission may add to the sound level in the receiving room so that the insulation value of the dividing panel may appear to be less than would be indicated by laboratory tests. Moreover, in buildings, the reverberant characteristics of the receiving rooms can vary considerably so that readings on this side of the panel can also vary. If the receiving room has a very short period of reverberation (as would be the case in a well-furnished living room) the reading will be lower, suggesting a higher insulation value. Additionally, the insulation characteristics of building panels vary to some extent according to their size.

When sound insulation values are given for practical use it is therefore usual to adjust experimental measurements to provide for conditions which are 'average'. For example, the sound insulation of a partition material would be given for a partition of average domestic or office size and as built into a stated general construction. Similarly, sound insulation values are 'normalised' to allow for a reverberation time of 0.5 s in the receiving room, this being an average for normally furnished and occupied domestic rooms.

Insulation varies with frequency

In practice we are never concerned with pure tones of one frequency but with speech, music or noise made up of a wide range of frequencies. Building materials, like all other materials, have varying insulation values at different frequencies. We therefore require to know how a material behaves at various frequencies in order to relate its insulation to the frequency spectrum of a given sound. For this reason the insulation properties of building elements are usually quoted for that range of frequencies which is important for noise control. Examples of such insulation values are plotted on the graph in figure 3.8 for three common materials. The general rise in insulation with frequency will be noted.

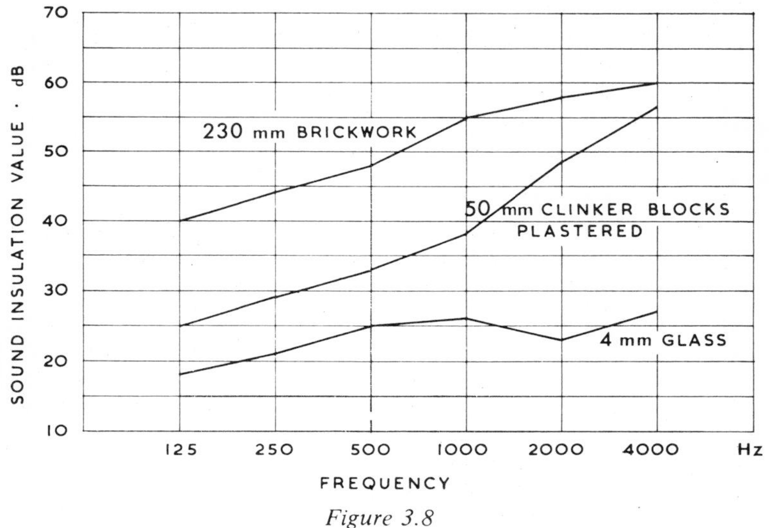

Figure 3.8

If the frequency spectrum of a noise is known, this can be compared with the insulating properties of the building element to discover whether it will be adequate at all frequencies to reduce the noise to an acceptable level. An example of this calculation is given on p. 125.

Average insulation values

The term 'average insulation value' is frequently used in describing such building elements as walls, partitions, floors and windows. Such values are obtained by averaging the results of laboratory or field tests over at least six frequencies (and more usually sixteen), the range being normally 100 to 3150 Hz in this country. Obviously, the more measurements that are taken, the more useful will the average figure be. Average figures are however sometimes deceptive because they can conceal weaknesses in insulation at certain frequencies due to the 'coincidence effect' referred to on p. 19.

Another difficulty sometimes arises when average figures from one laboratory are compared with those from another in which the *range* of test frequencies is different. This will affect the average figure and produce apparently different values for the same material.

Nevertheless, average figures are useful for an approximate comparison of one material with another and for preliminary calculations for noise control. They are used for this purpose in some of the calculations at the end of this section.

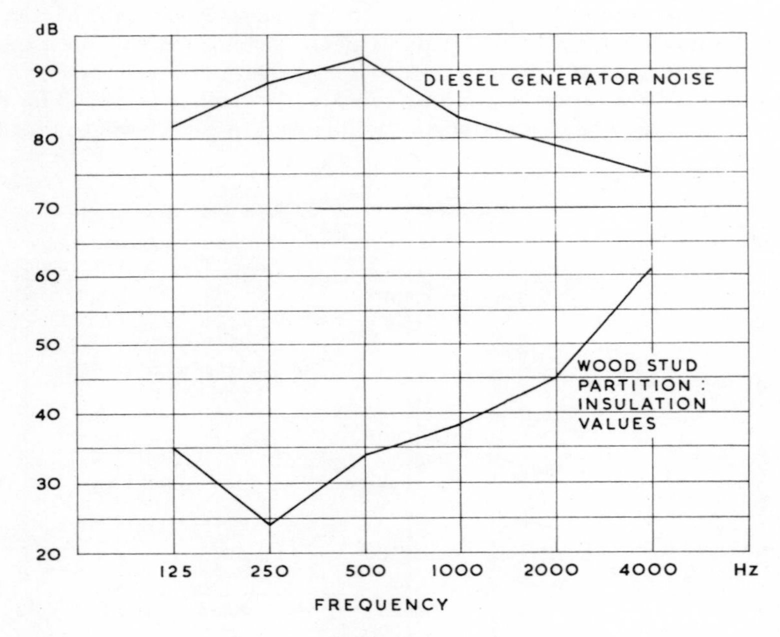

Figure 3.9

The most unreliable use of average insulation values can occur in those cases where a noise shows a marked increase of energy in part of its frequency spectrum, coinciding with a weakness in sound insulation at the same frequencies in a proposed partition. For example, the graph in figure 3.9 shows sound insulation values for a plastered, wood stud partition plotted against the frequency spectrum of the noise from a diesel generator. The gap between the two curves represents the levels to which the partition reduces the noise at various frequencies. At 2000 Hz the noise is reduced to 34 dB but at 250 Hz it is only reduced to 64 dB. Although the ear is less sensitive to the lower frequencies, this kind of discrepancy is excessive. Designers should therefore always obtain insulation values relative to octave frequency bands, and preferably third-octave bands. There is no reason why manufacturers of prefabricated partitions, double glazing systems, etc. should not supply such information since, in any case, an average figure must be derived from tests made in frequency bands.

Approximate average insulation values

We have seen in section 1 (pp. 18 to 22) that the insulation value of a homogeneous panel is determined by the interaction of a number of factors of which mass is only one. Nevertheless, mass is mainly responsible for the *general* rise in insulation as weight per unit area increases. We have seen that other factors, such as stiffness, give rise to resonances which tend to reduce insulation in certain frequency areas, so that average insulation values do not increase with mass as much as would be the case if the latter operated alone.

If the *average* insulation values of a range of building panels are plotted on a graph in relation to their weight per unit area, they will be found to lie fairly close to a curve rising 4 to 5 dB each time mass is doubled. This curve is shown in figure 3.10. This characteristic is useful in those cases where there are no published data for a proposed construction. If its weight per square metre is calculated, its approximate average insulation value can be assessed from the graph.

It must, however, be emphasised that the true insulation value could be more *or less* than that indicated by as much as 3 dB and that in any case average figures can conceal weaknesses in insulation in certain parts of the frequency range. It may, however, be observed that to increase, for example, the average insulation value of a 225 mm brick wall by only 5 dB its thickness has to be doubled. From a practical point of view this sets a severe limit on the improvement of sound insulation by weight alone. Other methods of improving sound insulation then become necessary and some of these will be described below.

Double-leaf construction

If, instead of dividing two rooms by a single homogeneous partition, the same amount of material were employed in two leaves with a cavity between, it would be expected that each leaf, being half the weight, would have an *average* insulation about 4 to 5 dB below that of the single wall. Thus if the original partition had an insulation value of 40 dB then each leaf of the double partition should have an insulation capability of 35 dB. In most cases this would be about right. One would further expect that, since there is a transmission loss of 35 dB as between the first room and the air cavity, and then a further loss of 35 dB between the cavity and the adjoining room, the resulting difference between the two rooms would be 70 dB.

If the cavity were very wide and there were no connections between the two leaves, even at the edges, this also would be approximately true. In practical building construction the above requirements cannot of course be met, with the result that increases in insulation such as the above cannot even be approached and in some cases improvement is very small indeed.

Because the cavity has to be limited in width and closed at the edges, the trapped pocket of air couples the two leaves rather like a spring, conveying

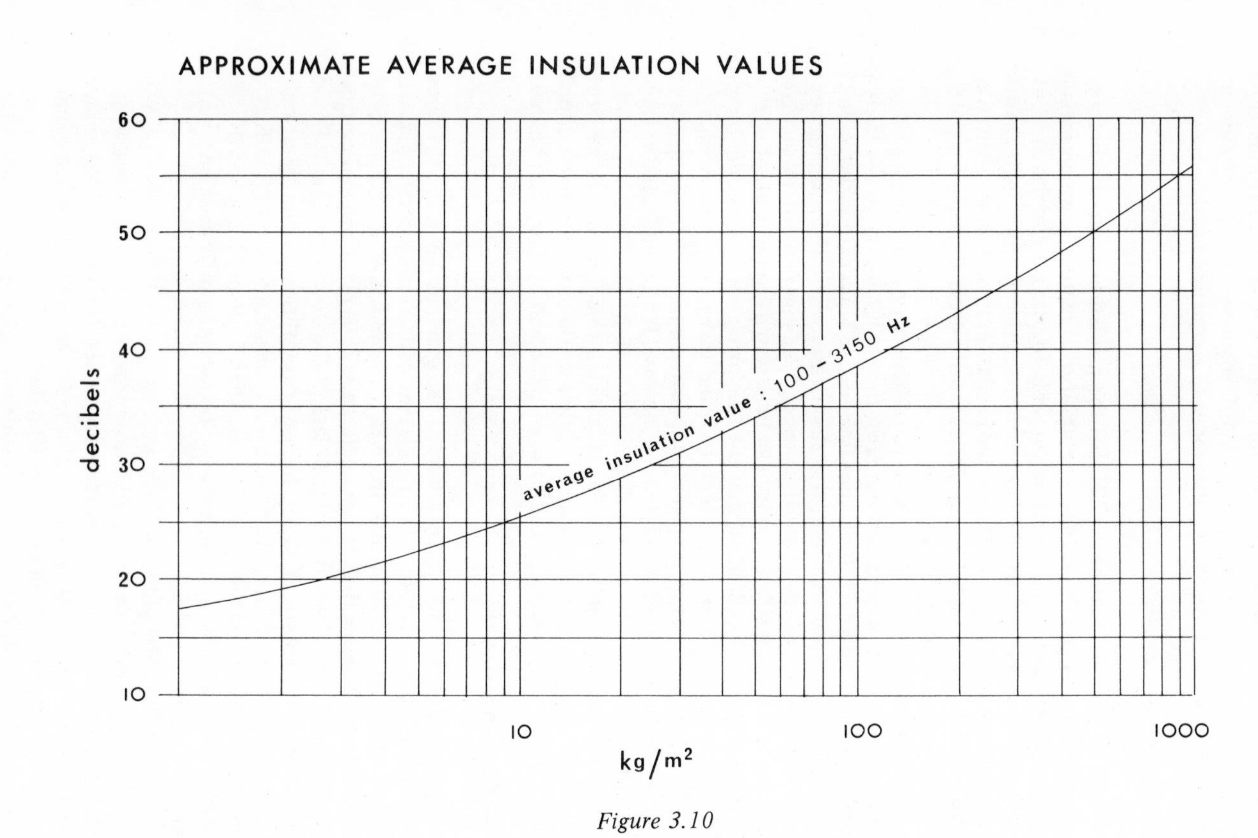
APPROXIMATE AVERAGE INSULATION VALUES
decibels
60
50
40
30
20
10
average insulation value : 100 – 3150 Hz
10
100
1000
kg/m²

Figure 3.10

vibrations effectively from one leaf to the other. Any connections between the two leaves will also convey vibrations directly from one leaf to the other, particularly if such connections are distributed over the area of the partitions, such as in the case of the wood studs in a stud and plaster partition or the metal ties in a brick cavity wall. There will also be cavity resonances which tend to reduce insulation, though these can usually be absorbed by the introduction of sound-absorbing material in the cavity, or if the inside surfaces of the cavity are sound-absorbing. Finally, the division of the material into two thinner leaves shifts the coincidence effect to another part of the frequency spectrum where the loss of insulation may be more serious, apart from the fact that the 'coincidence dip' is usually deeper.

Prediction of the net results in terms of insulation for new forms of cavity construction is therefore extremely difficult and the designer must consult or obtain actual field test figures, *versus frequency* before assuming that the introduction of a cavity will necessarily show a marked improvement over a solid construction.

In general, however, it may be said that for low frequency insulation adequate mass and a sufficiently wide cavity are important. The dip in the insulation curve due to 'coincidence' can be reduced by dissimilar thickness of leaves. Absorbent material in the cavity reduces the effect of cavity resonance and bridging the cavity must be avoided if at all possible.

Effect of doors, windows and rooflights

We have so far considered the case of walls, partitions and roofs of simple uniform construction, whereas in practice walls have windows, partitions have doors and roofs have rooflights. If we consider the case of an external wall with a closed window and an external noise source, then the sound level close to the window will be mainly determined by the insulation value of the glazing. The *average* noise level in the room will, however, be determined by the wall and window acting in combination.

A method of calcualting what is called the 'net insulation value' of two such elements is described on p. 122, but it may here be stated that when a window is introduced into a wall the net insulation of wall and window is usually closer to that of the window than of the wall. The loss of insulation caused by the window is greater than would be expected from a consideration of its relative size.

Figure 3.11 demonstrates this in the case of five examples of an external wall with different areas of glazing. It is assumed here that the wall has a typical insulation value of 50 dB and that the window *when closed* provides an insulation of 20 dB. The *net* insulation values are given under each drawing.

This disproportionate reduction in insulation occurs in all cases where an acoustically 'weak' element is introduced into a potential sound barrier, so that it applies not only to windows, doors and rooflights but wherever a change of material or thickness occurs which causes a reduction of insulation.

Although it has been said that the net insulation of an external wall is usually determined more by the insulation value of its windows than of the wall itself, it

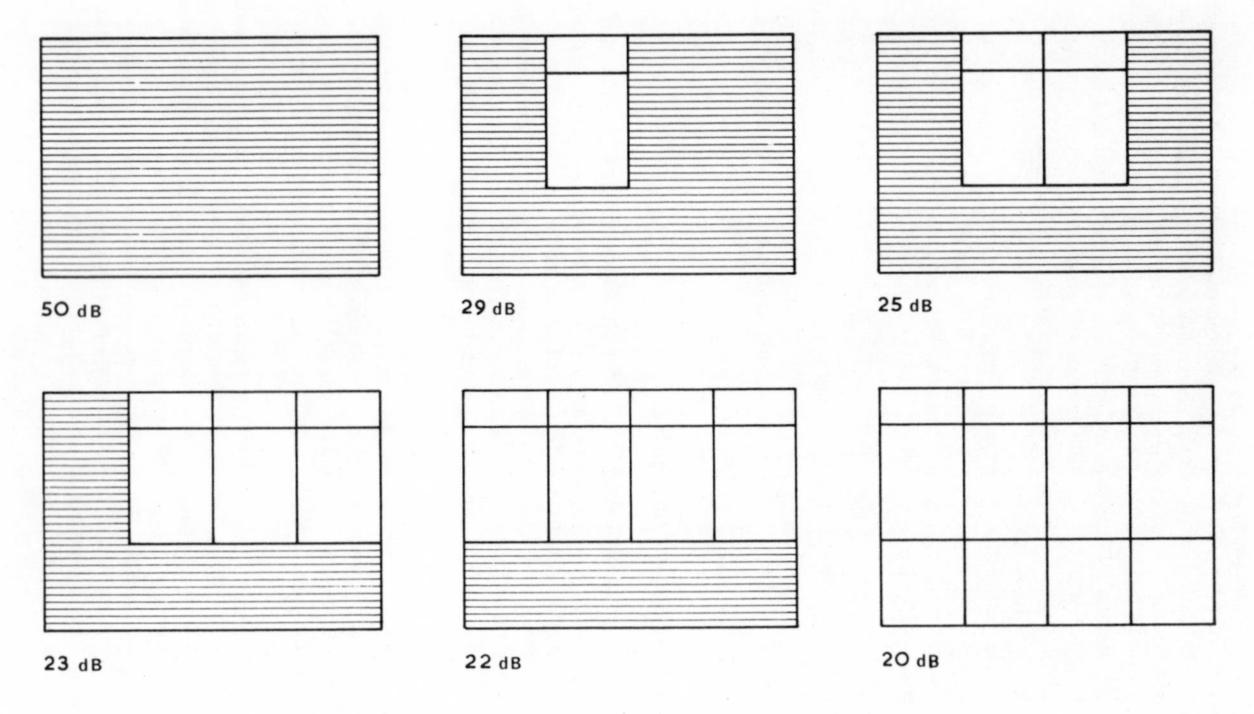

Figure 3.11

should not be inferred that the recent tendency to increase window sizes is of no great importance. The reductions in insulation as window proportions increase in figure 3.11 are seen to be significant when it is remembered that an increase of 10 dB is subjectively 'twice as loud'.

Loss of insulation due to voids

When windows or doors are opened insulation can of course become negligible. To a person standing near an open window insulation is nil. In other parts of the room insulation will depend on the proportion of opening and the relative positions of listener and void. In domestic rooms with windows partially open the *average* insulation is likely to be of the order of 5 dB—very little defence against modern traffic noise.

A reduction in insulation can also occur if windows (or doors) are not air-tight when closed. The loss of insulation due to the free air paths of even small gaps or cracks is considerable. For this reason insulation values are often given for windows 'normally closed' and windows 'sealed tight' or for doors 'normally hung' and those whose junctions are 'sealed' by resilient beads. For a sound barrier to be fully effective it must be 'complete'. Partitions must fit tightly within the structural surround; sliding, folding screens should be air-tight and vents in external walls must be excluded unless they are acoustically baffled.

Noise control and ventilation

Closed windows imply the need for alternative means of ventilation. Here it is possible to exchange one kind of void for another. If simple ventilation grilles are inserted in the wall facing the source of noise, one might as well open the windows. Such vents can, however, be 'baffled' to reduce the ingress of noise, as shown in figure 3.12A. The ventilation opening is 'staggered' to check direct sound transmission and to provide an adequate length of sound-absorbing lining to the duct. The duct can of course be staggered on plan. Its effectiveness will depend on the coefficient of absorption of the lining material, its length and the ratio of perimeter to cross-sectional area. In these terms attenuation of sound can be calculated. Where fans are associated with such vents they should be located at the external end of the duct because fans are themselves a source of noise.

Ventilation ducts employed in centralised systems of ventilation can result in the transmission of noise from one part of a building to another, in addition to allowing the ingress of external noise. Figure 3.12B shows how a shared ventilation system can result in the exchange of noise between rooms, irrespective of the direction of air movement. To prevent this the whole, or a sufficient length, of the ducts between rooms should be lined with sound-absorbing material.

Such systems present an additional problem—the transmission of ventilation plant noise and external noise into the ventilated rooms. The architect should threfore satisfy himself that the heating and ventilation specialist has taken adequate precautions to reduce duct noise to an acceptable level. The maximum allowable noise level should in fact be specified to suit the use of the accom-

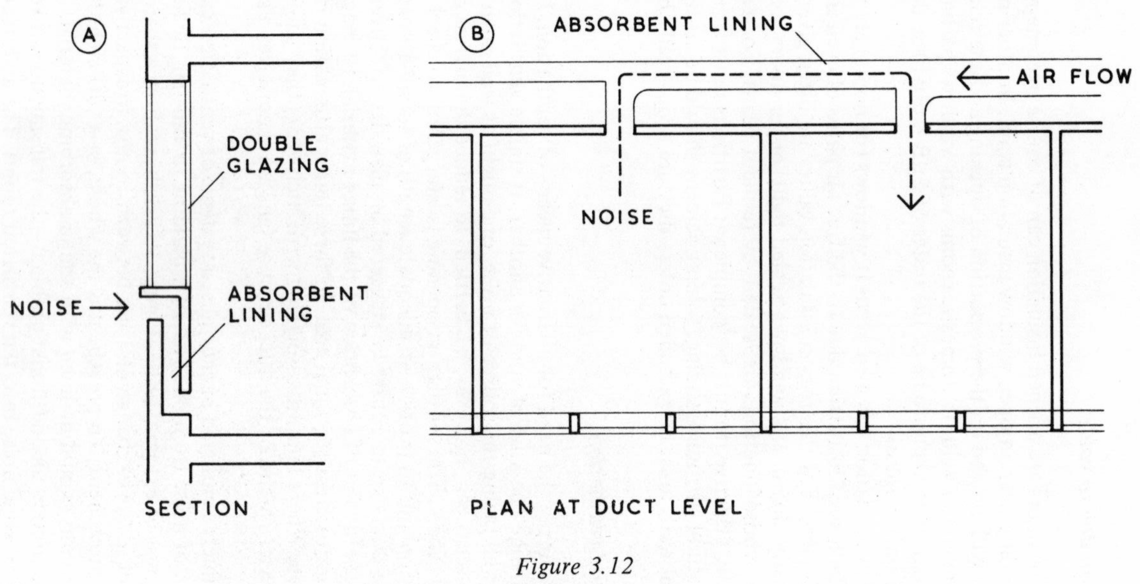
A
DOUBLE
GLAZING
NOISE
ABSORBENT
LINING
SECTION
B
ABSORBENT LINING
AIR FLOW
NOISE
PLAN AT DUCT LEVEL

Figure 3.12

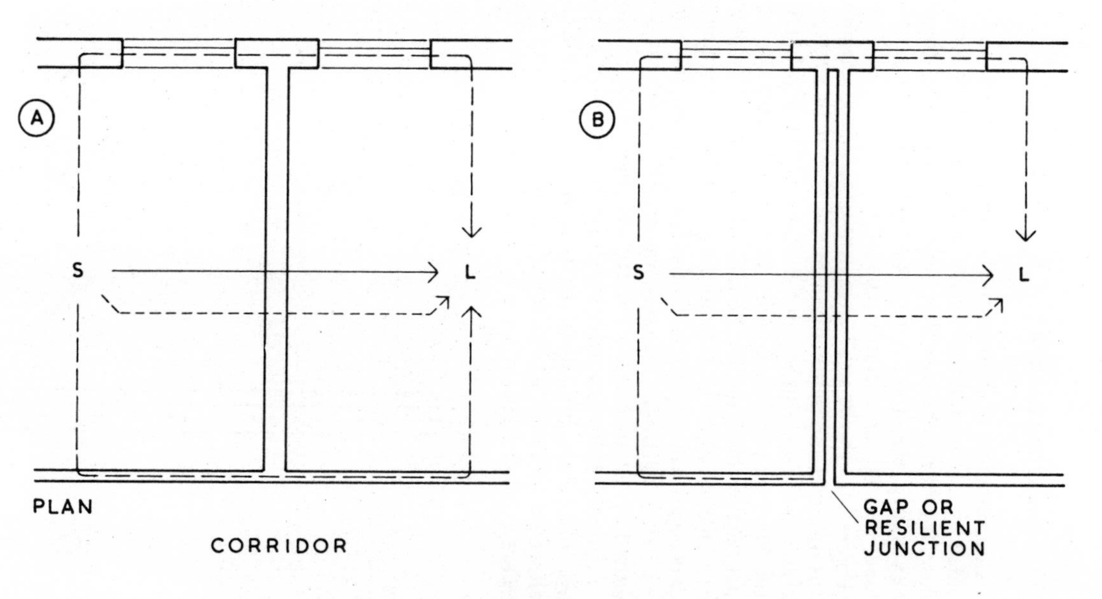

Figure 3.13

modation (see p. 112). In brief, such noise reduction will involve the resilent mounting of ventilation plant and the absorption of fan noise, air-borne equipment noise and possibly external noise by an 'attenuator' between plant room and the system of ducts. Metal ducting can itself carry sonic vibrations but these can be absorbed by a length of flexible duct near the point of origin.

Indirect sound transmission

The disadvantages of indirect or flanking transmission of sound can often be avoided by minor modifications in design. as will be seen in the examples that follow. It will also be seen that, unless all possible paths of sound transmission are examined, the effectiveness of a sound-insulating barrier between source and listener may be considerably reduced.

Figure 3.13 shows an arrangement of two rooms leading off a corridor. The paths of direct and indirect transmission are shown by arrows, the dotted line representing transmission through floor and ceiling. In (A) the corridor partition is the 'weakest path' for sound transmission, being considerably lighter than any other part of the structure. Sound will, as it were, by-pass the heavy dividing partition and reduce its effectiveness. In (B) the dividing partition is of two independent leaves and there is a gap or resilient joint where the two leaves meet the corridor partition. Flanking transmission along the corridor partition has been checked by 'acoustic discontinuity' and the double partition will be fully effective.

Figure 3.14 shows a party structure between two dwellings constructed of two leaves of 115 mm brickwork. In itself the average insulation value of the party wall would be about 55 dB. Sound, both air-borne and impact vibration, will, however, follow indirect paths as shown and the *effective* sound insulation is

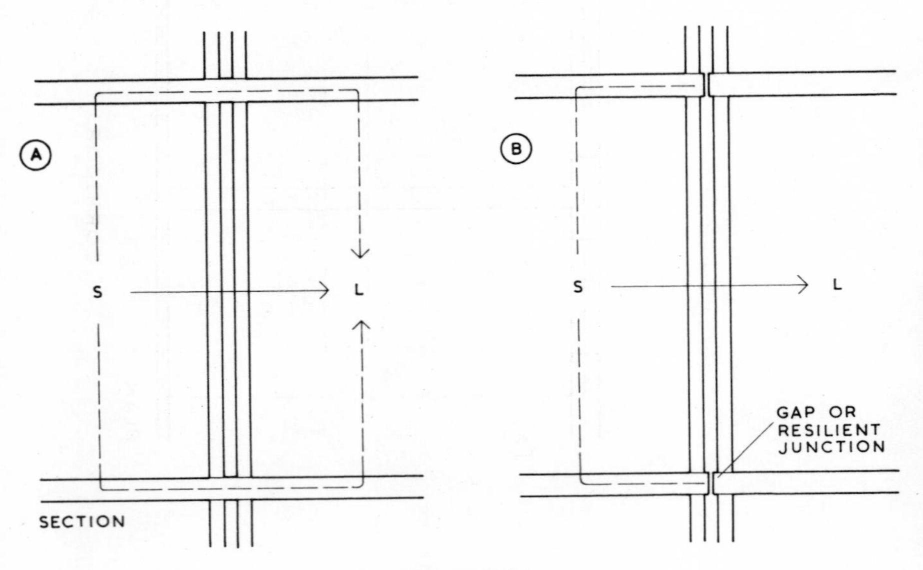

Figure 3.14

likely to be considerably reduced. On the other hand, if floor transmission can be checked by discontinuity, as in (B), the double-leaf wall can realise its full insulation value.

In figure 3.15 two examples are shown of a partition dividing offices. In (A) the insulation between the rooms will be determined by the insulation value of the partition if, as is likely, the floor construction is heavier. For example, a well-designed sectional office partition could provide an average insulation of about 30 dB. In (B), however, the same partition is stopped short at the level of a perforated, suspended acoustic ceiling. Sound by-passes the partition, as shown, and insulation is considerably reduced.

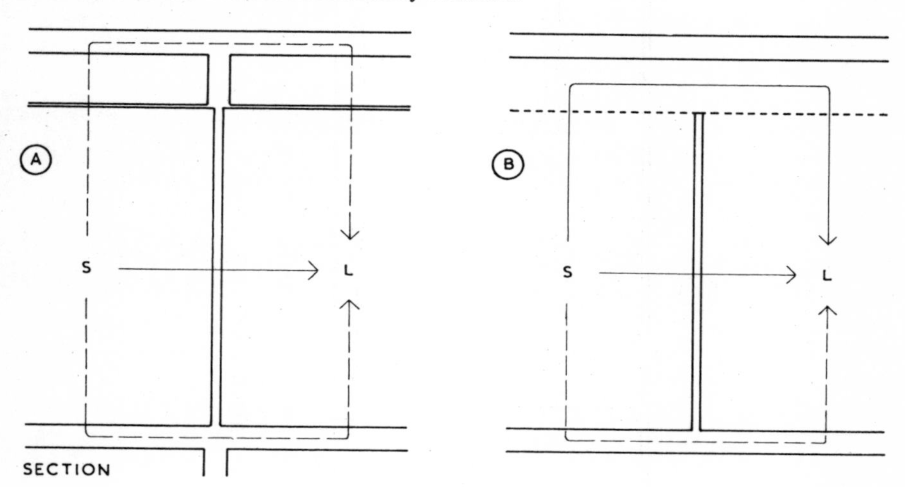

Figure 3.15

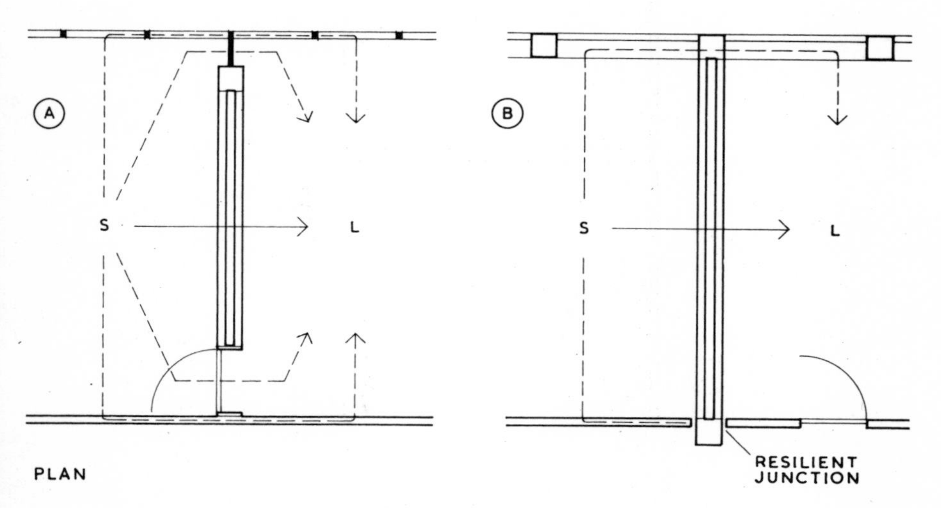

Figure 3.16

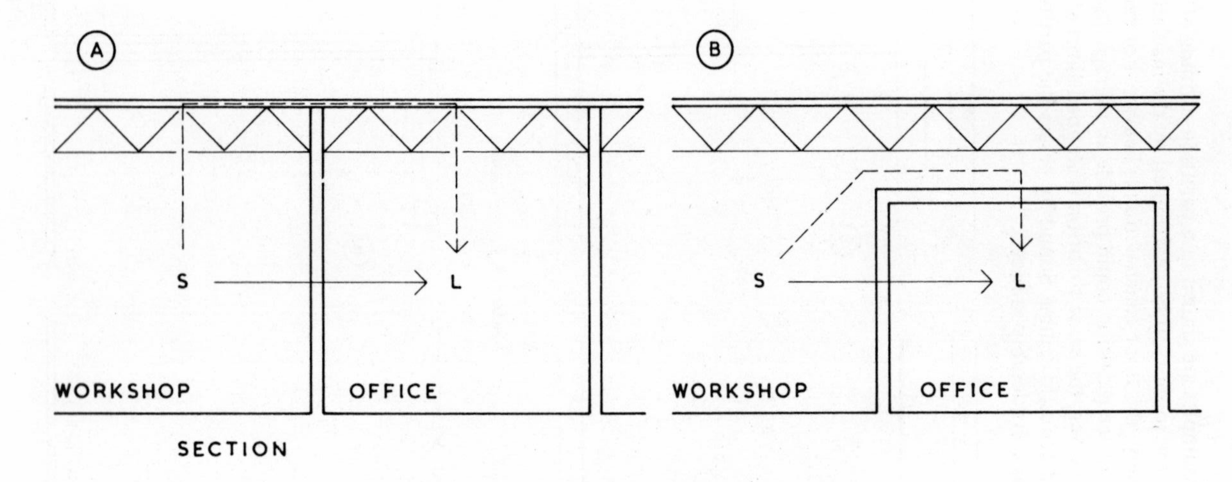

Figure 3.17

Figure 3.16 shows a typical cantilevered construction with light external cladding of the 'curtain wall' type (A). The arrows show the paths of sound transmission between the two rooms and the ineffectiveness of the double partition is clear. The more robust external wall construction in (B) and the avoidance of 'weak' elements in the dividing partition will make its double-leaf construction worth while.

Figure 3.17 shows how noise may be indirectly transmitted from one internal space to another by way of light roof decking. In (A) what may be a sufficient sound barrier has been placed between workshop and office but this can easily be by-passed by sound transmission along the roof deck. Where an office has to be provided in a workshop it may be more practical acoustically to build a separate enclosure, as in (B), and ensure that all enclosing walls *and its roof* have the required insulation value.

Fully discontinuous construction

As explained on p. 69, there are practical limits to the insulation that can be provided by normal constructional methods. They are in fact in the region of average insulation values of 50 to 55 dB. Some sound insulation problems require a reduction greater than this and here the principle of discontinuity must be fully exploited.

In principle, the method is to completely separate those surfaces which receive the sound from those surfaces which surround the listener. This may be done in two situations, shown diagrammatically in figure 3.18. In (A) the room containing the listener is built as an acoustically independent 'box', independent, that is, of the rest of the structure. External noise impinging on the structure can only to a very limited extent reach the surfaces surrounding the listener because all paths of sound transmission are broken, either by an air space or by resilient material. The sound radiated to the listener by the surfaces which surround him is thereby considerably reduced. In this way, average sound reduction values of 60 to 65 dB can be achieved without recourse to massive construction.

Drawing (B) in figure 3.18 shows the alternative situation in which the source of sound is enclosed by an acoustically independent 'box' so that the transmission of sound to other parts of the building is reduced by complete discontinuity.

The important factor in this type of construction is that acoustic separation must be complete. At no point should the 'gap' be bridged by rigid material. Problems immediately arise from the need to gain access to the enclosure and from the need to provide light, ventilation and services. Complicated construction results and the need for careful detailing and site supervision is paramount. The points at which the 'gap' may be bridged are indicated by arrows in figure 3.19, which takes a relatively simple case to show where acoustic discontinuity must be maintained by resilient junctions.

The same principle of discontinuity is employed on a larger scale in the separation of whole buildings or sections of buildings in which one part produces noise and the other part is required to be protected from it. We are often in this case not only concerned with air-borne noise but also with the greater problem of checking the transmission of impact, or contact, vibrations which, if allowed

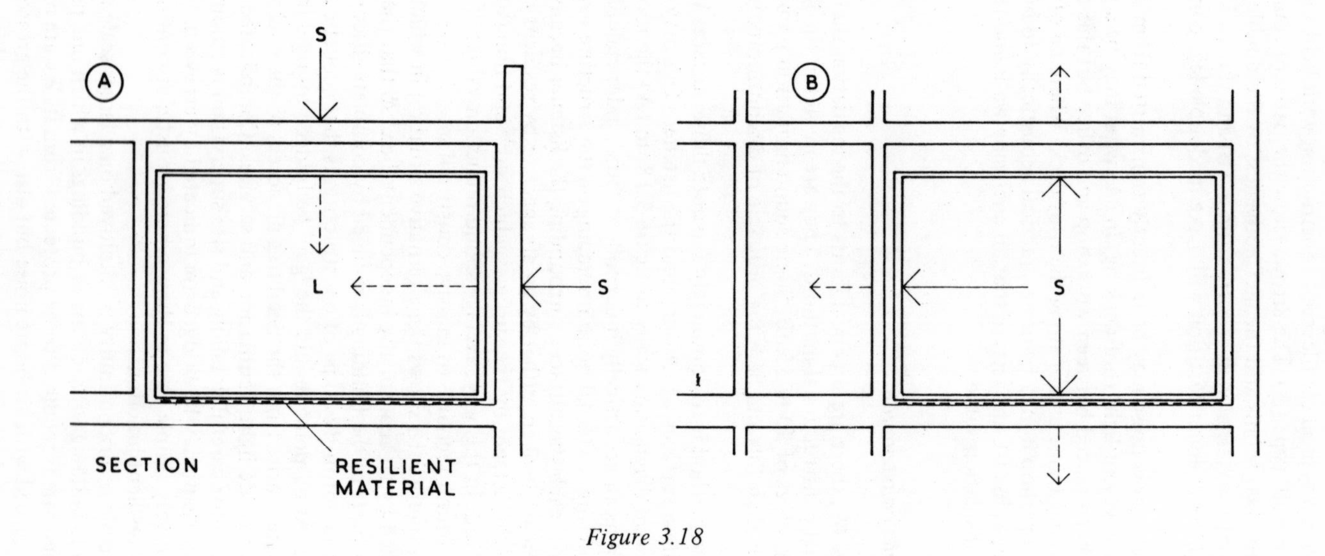

Figure 3.18

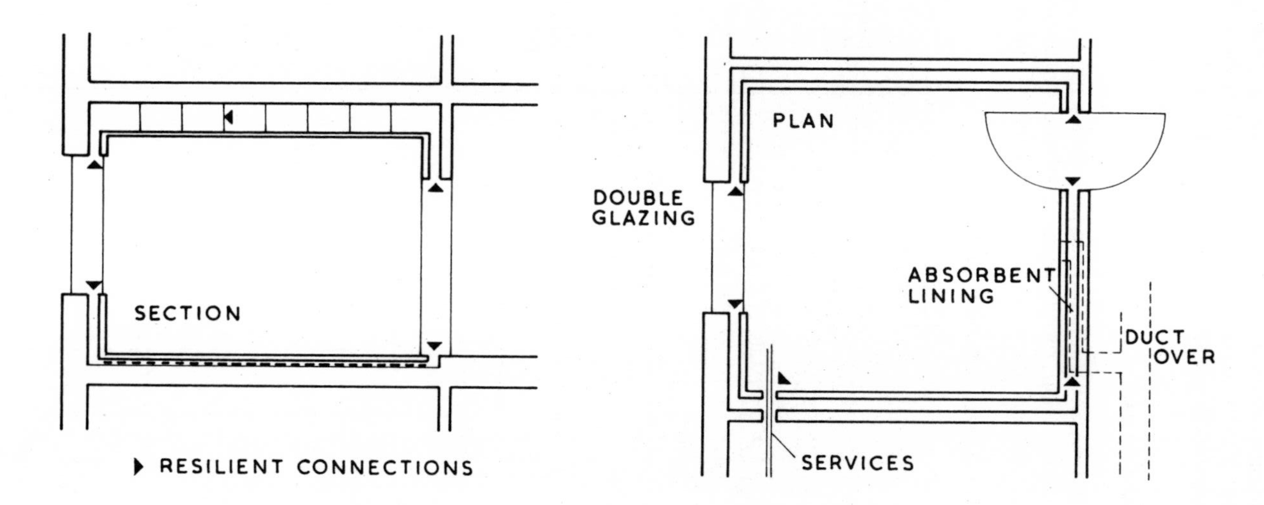
SECTION
RESILIENT CONNECTIONS
PLAN
DOUBLE GLAZING
ABSORBENT LINING
DUCT OVER
SERVICES

Figure 3.19

to travel through the structure, will produce air-borne noise in remote parts of the building. A common case is the association of offices with factory workshops in one building group. If the two parts share a common structure the transmission of vibrations between workshop and offices is difficult to reduce.

Design of windows and doors

In most buildings windows and doors provide the 'weak path' for air-borne sound transmission. In this connection, rooflights can be thought of as horizontal windows, equally vulnerable to external noise, particularly from aircraft.

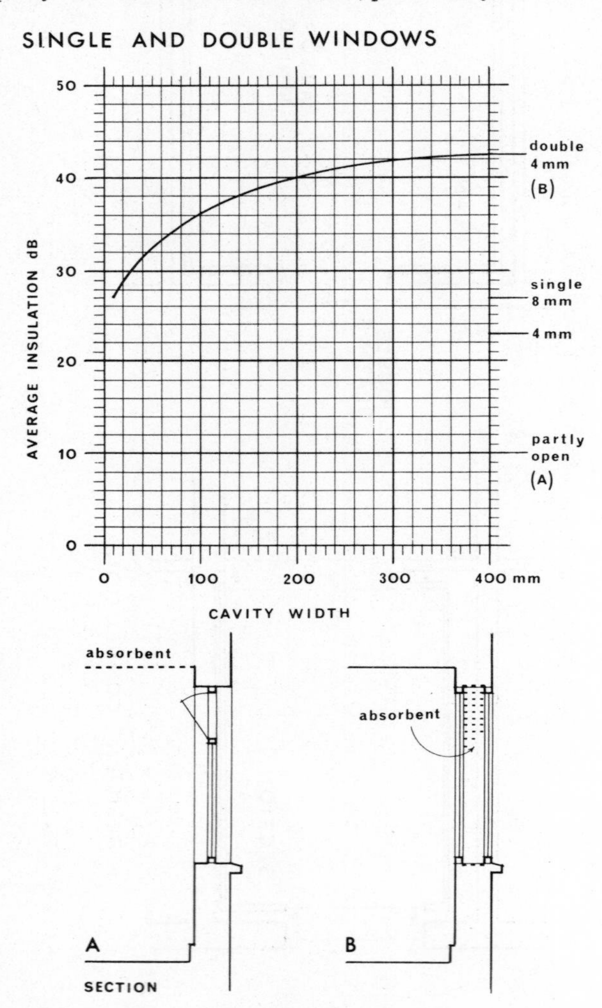

Figure 3.20

We may begin by examining the open window. In domestic and office buildings the external wall and window is unlikely to provide an average insulation of more than 10 dB when the window is partially open for ventilation. 'Average' here means average over the whole room. Near the open window insulation will be much less. Some improvement can be obtained by the use of a sound-absorbing ceiling associated with a bottom-hung ventilator as shown in figure 3.20A. In most cases of road traffic noise the reduction will, however, be too small to provide comfortable conditions in rooms of low background noise level. Fixed or double glazing then becomes necessary, with attendant problems of ventilation.

Fixed, single glazing provides a considerably more effective barrier to noise than openable windows. Average reductions of 23 to 25 dB are attainable, which will deal with many of the less severe problems encountered in practice. To achieve this degree of insulation the window must however be air-tight, insulation then depending on the weight of the glass.

The most effective form of window insulation is that provided by fixed double glazing, by which average insulation can be raised to 40 dB or more. To attain this reduction, however, the following provisions are necessary

(1) the space between the glass should be at least 200 mm,
(2) both areas of glazing must be air-tight and
(3) the head, sill and reveals should be lined with an efficient sound-absorbing material.*

Figure 3.20B shows such a window and a graph which relates average insulation to the width between the two panes of glass. It will be seen that, if the gap between the glass is only 10 mm the insulation is no greater than if the two panes of 4 mm glass were combined in one pane of 8 mm thickness. Moreover, the presence of the sound-absorbing lining adds to the insulation if space for this is allowed. For example, in the case of a double window with a 100 mm cavity, the sound-absorbent material adds 4 dB to over-all insulation.

Where glazing is made air-tight the simplest form of alternative ventilation is to build a fan into the external wall. Whether the air is extracted or drawn in, noise will penetrate unless a baffled opening is provided, lined with sound-absorbing material. The fan should be located at the external end of this opening so that both external noise and fan noise are reduced by the absorbent material (see figure 3.12).

The principles set out above will of course apply equally to rooflights, in which a fixed rooflight and laylight can form the equivalent of a double window. Additionally, rooflights can be screened from street noise by parapets.

Doors provide a similar 'weak path' in defence against noise because they are normally much lighter than the surrounding wall or partition and because they are seldom air-tight. It follows therefore that they should be made as heavy as possible and should close on resilient beads.

*The improvement in insulation due to the introduction of sound-absorbing material will depend on cavity width and other factors in the design of the window. An improvement of 2 to 4 dB in average insulation value is possible.

Design of sound lobbies

However heavy and tightly fitting a door may be, it is required to stand open periodically and then provides no barrier to the transmission of noise. For this reason 'sound lobbies' are provided, for example, between a foyer and an auditorium, on the assumption that both sets of doors will infrequently be open at the same time during a performance. When both pairs of doors are closed the total insulation is likely to be comparable with that of the auditorium wall.

To be most effective, however, in normal use, the lobby should be lined with sound-absorbing material and the doors made as heavy as practicable (figure 3.21A). When both pairs of doors are closed an over-all average insulation of 45 dB is then possible. In larger lobbies the staggered arrangement of doors shown in figure 3.21B will reduce direct sound transmission when both sets of doors are open.

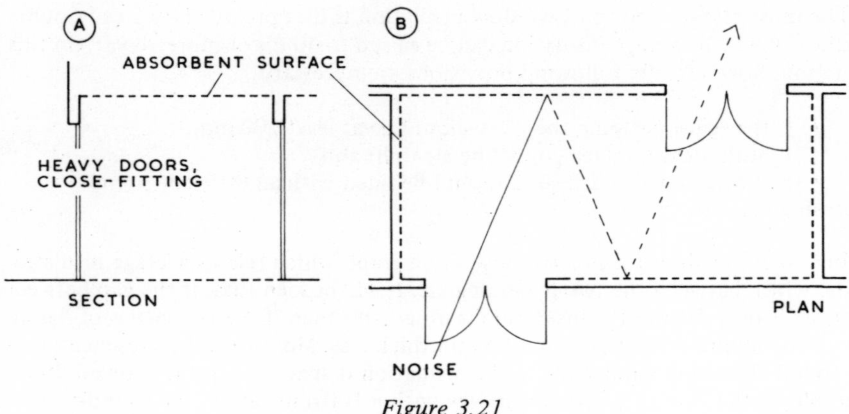

Figure 3.21

Ground-borne vibrations

We have seen that the transmission of air-borne sound through a building structure is largely dependent on the weight of construction, and this applies to suspended ground floors as shown in figure 3.22A. Where, however, the floor is laid direct on the ground (as in the case of site concrete) transmission will be negligible (figure 3.22B). If, however, a vibrating element, for example a machine, is in direct contact with such a floor, vibrations may be sufficiently powerful to travel through the concrete slab and cause significant structural transmission. It is therefore always advisable to isolate such elements from the floor slab by means of resilient mountings.

Ground vibrations, sufficient to cause troublesome sound transmission, can also be caused by surface or underground rail traffic. An example is shown in figure 3.23A, in which such vibrations are transmitted to a building structure, resulting in the emission of air-borne sound within the building, even though, as in the example, the occupants of the building are shielded from air-borne sound by the embankment, boundary wall and the external wall of the building.

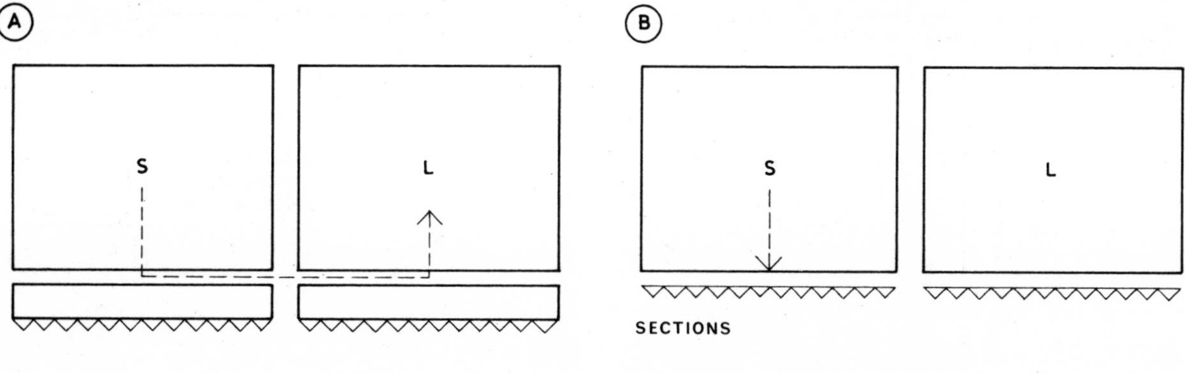

Figure 3.22

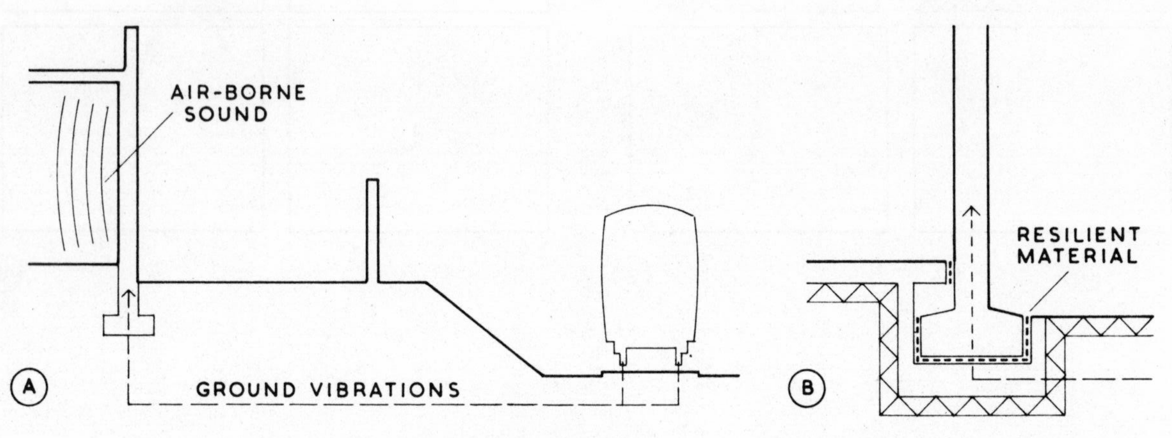

Figure 3.23

In such situations it may be necessary to isolate the building structure from the ground by means of double foundations, with a vibration-damping material between the main and sub-foundations as in figure 3.23B. In some cases it may be more expedient to deal with the problem by means of the discontinuous construction illustrated in figure 3.18.

Reduction of noise by distance

The reduction of sound as distance increases has been explained in section 1 and reference should be made to the graph on p. 13. It will be seen that, in a free field, sound level reduces by 6 dB each time distance from the source is doubled, in the case of a *'point source of sound'*. This applies, for example, in the case of individual road vehicles and aircraft if the first measurement is taken at a distance considerably greater than the size of the sound source.

A stream of road vehicles cannot, however, be considered as a point source of sound, nor is it a true 'line source', since the emission of noise from each vehicle varies considerably and they are seldom equally spaced. It has been found, as would be expected, that the average rate of reduction of noise in this situation lies somewhere between that of a point source and a line source, in fact about 4 dB each time distance is doubled. This rate of reduction will however apply only beyond a distance of about 15 metres from the road.

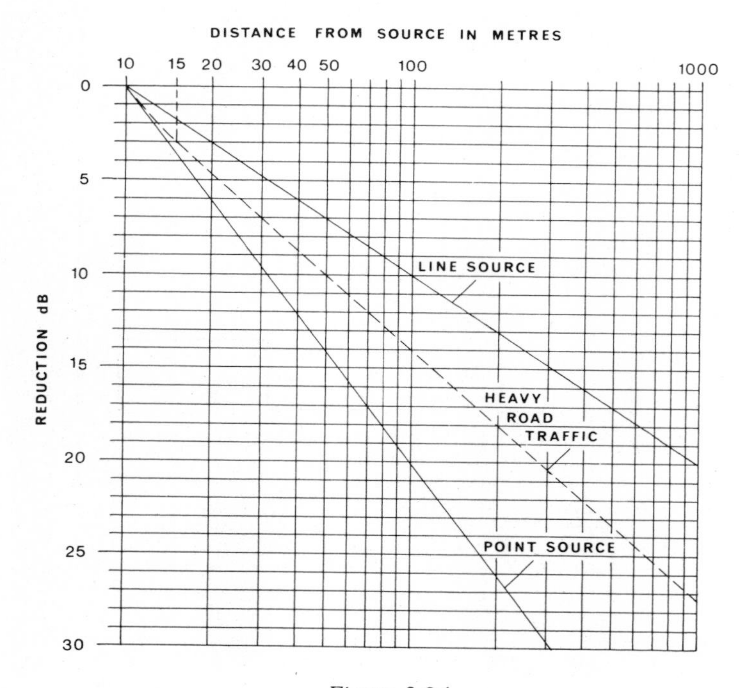

Figure 3.24

The graph, figure 3.24, shows this rate of reduction compared with that from a point source and it will be seen that, closer to the road than 15 m, the rate of reduction of noise from a stream of vehicles approaches the inverse square law, since individual vehicles are tending to dominate the situation.

Noise level contours

It will be seen from the above that, if it is required to establish the noise levels on a building site adjoining a busy main road, it is only necessary to measure the traffic noise at one point on the site and then from this can be extrapolated the expected levels at other points, having regard to the effect of distance only.

This has led to a technique for 'contouring' sites with noise levels, much as a surveyor might contour a site with physical levels. The advantage of being able to do this is evident when the siting of buildings is being considered. In preparing such contours other factors than distance have to be taken into account and the more important of these will be referred to below. Figure 3.25, however,

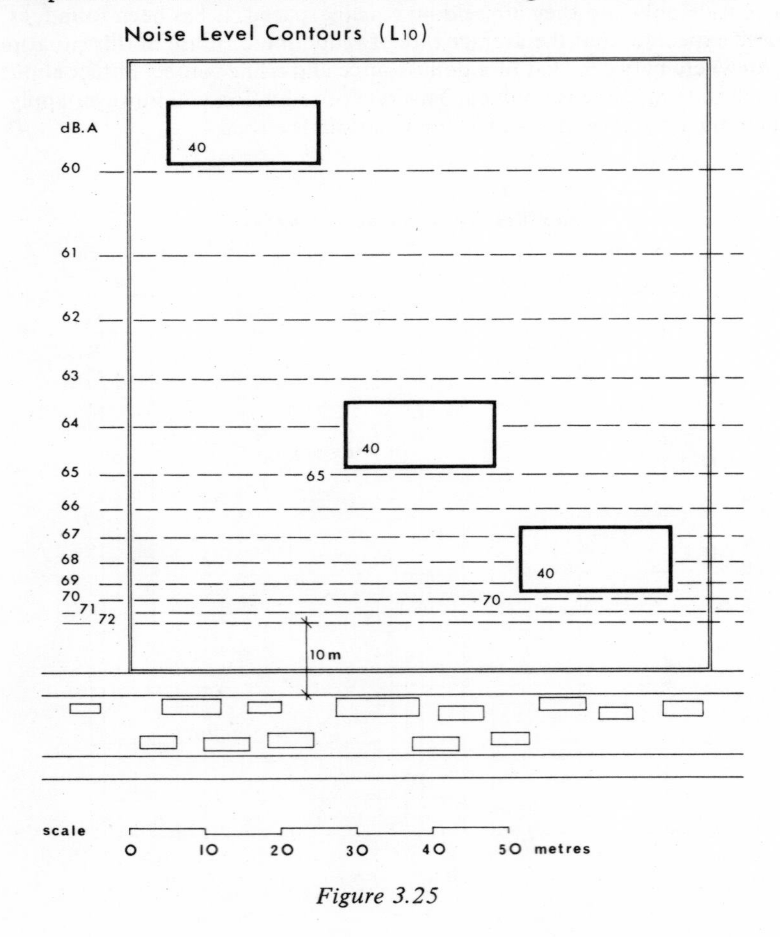

Figure 3.25

illustrates the use of noise level contours in design. Here, alternative positions for a building are being considered and it will have bedrooms facing the road. The contours immediately reveal the problem of defending these rooms from excessive noise. By deducting an acceptable intrusive noise level for bedrooms (say 40 dB.A) from the level of external noise in each case, the required average insulation values for the external walls and windows can be estimated.

Road traffic noise is, however, a fluctuating source of sound and before sound level contours can be drawn a decision has to be made as to a reasonable level to adopt as a basis for the calculation, that is, the level for the first contour.

It has been found that the sound level exceeded for 10% of the time (over an extended period) correlates well with the degree of noise disturbance to be expected. In other words, if at 15 m from the road noise exceeds, say, 85 dB.A for 10% of the time then buildings on an adjoining site should be adequately insulated from noise of this level, or from such reducing noise levels as will be indicated by the contours.

The period of time for taking such readings is from 06.00 to 24.00 hours on a normal working day and the resulting noise level is then described as L_{10} (18 h). It is also possible to estimate L_{10} from a knowledge of traffic density, types of vehicle, average speed and road gradient. A detailed treatment of the subject is given in booklets published by the Department of the Environment.

Ground absorption and reflection

When sound waves travel over an absorbent surface, such as grass, their energy near ground level is reduced, and this absorption is progressive. Conversely, if sound travels over a hard surface, such as paving, then by reflection sound level may be increased. This increase is not progressive and if the first meter reading includes ground reflection no further increase will occur.

These factors have to be taken into account in producing sound level contours and, since ground effects will depend on the average height of sound propagation, contours should indicate the relevant heights of sound source and reception point, possibly by means of a section.

It follows therefore that, as far as ground floor accommodation is concerned, an expanse of grass or other planting between the source of sound and the building can be advantageous. An area of hard paving, on the other hand, can raise noise levels by about two decibels. Having in mind that in this case the benefit of ground absorption has been lost, a very useful reduction in noise has been given away. A common example of this is the siting of a school playground between road and classrooms.

As with the effect of distance, however, the reduction of noise due to ground absorption can be over-estimated if reference to data is not made and, in any case, a material advantage is only to be expected for ground floor accommodation.

Reduction of external noise by screening

If a screen wall is set between a source of sound and a listening position, the level of sound heard will depend on the amount of sound energy diffracted over the top edge (assuming that the wall is of sufficient length for 'end diffraction' to be ignored).

As we have seen in section 1 (p. 30), more diffraction occurs at low frequencies than high, so that the frequency spectrum of a noise is also modified. Since the ear is more sensitive to high frequencies than low, this characteristic is an advantage in the reduction of the *loudness* of the noise heard..

A boundary screen of sufficient length can therefore be usefully employed to reduce road or rail traffic noise in respect of ground floor accommodation, and possibly at first floor, if within the 'sound shadow'. The screen can be a wall or fence, provided the latter is impervious to the passage of direct sound. An embankment or similar change in ground level can act as a screen, and buildings which are not vulnerable to noise disturbance can be used to screen more sensitive accommodation.

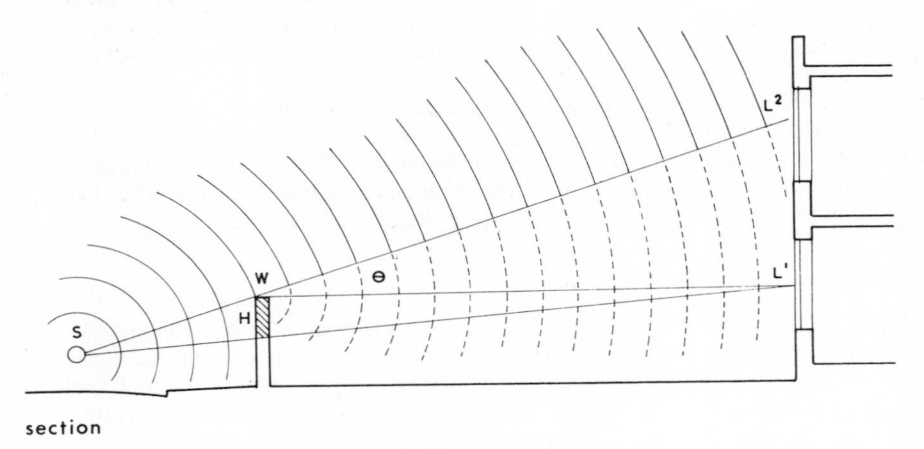

Figure 3.26

The phenomenon is illustrated in figure 3.26, which also indicates the main factors that determine the reduction of noise at the reception point. At L_1 a listener hears sound waves considerably weakened by diffraction, whereas at L_2 sound intensity is only very slightly reduced. It should be noted that, since the energy in the diffracted waves has been extracted from the uninterrupted waves, there will be a small reduction in intensity even outside the sound shadow, as shown by figure 3.38.

The reduction in sound level between source and reception point will depend on

(1) the total distance travelled by the uninterrupted and diffracted sound waves (SWL)
(2) the projection of the screen into the direct sound path (H)
(3) the angle of diffraction (θ)
(4) the frequency spectrum of the noise.

Examples of the calculation for the effect of screening are given later in this section.

Screening to reduce noise can of course be employed in other situations, some

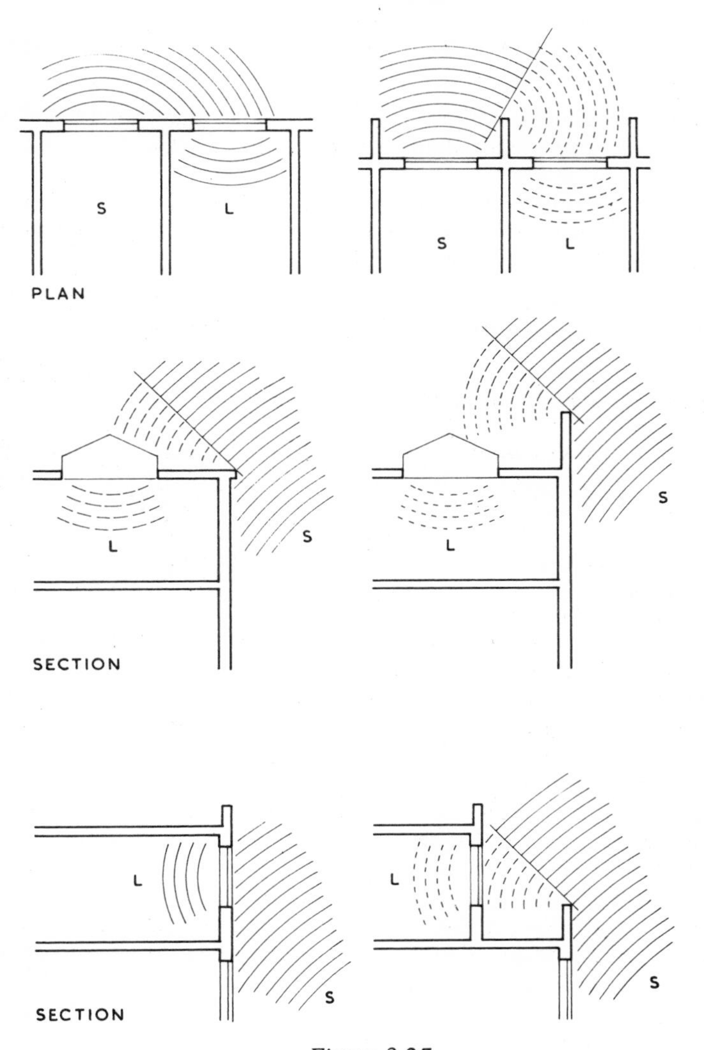

Figure 3.27

of which are shown diagrammatically in figure 3.27. It is important, however, to make an approximate calculation of the effect of screening in each situation. The reduction of noise may be only marginal and, in itself, not provide a complete solution.

Screening by planting

The planting of trees between a noisy road and buildings is generally thought to provide an effective screen. In most cases this belief is over-optimistic. While it is true that everything helps in the battle against noise, the following reservations must be made

(1) Sound reduction is due mainly to foliage so that sound travelling near ground level will not be reduced unless tree planting is associated with hedges or shrubs.

(2) Sound reduction will only be significant when trees are in leaf.

(3) The reduction of high noise levels will be relatively small unless a very wide belt of trees is planted (see p. 116).

Nevertheless, foliage absorbs sound more effectively at high frequencies so that at moderate sound levels the reduction may be subjectively significant.

Planning and the analysis of the problem

When the principles discussed above have been understood the architect is in a position to reduce the problem of noise control by both site planning and internal planning. It is for this reason that general considerations of planning have been left to the end, although in design they would take first place.

In what follows it is not overlooked that measures taken to reduce a noise problem by planning may conflict with requirements of circulation, aspect, prospect and the many other factors which have to be taken into account in the design of a building. Here, the kind of value judgement with which the architect is all too familiar will often occur, not the least of these being associated with questions of cost. A simple example of such a conflict of requirements is illustrated by the example on p. 115 in which the decision has to be made between facing classrooms towards the south and the noise of traffic, or towards the north and away from noise.

The control of noise really commences with the choice of site and the importance of noise as a factor in that choice will of course depend on the purpose of the building. Reference to the table of acceptable intrusive noise levels on p. 112 will assist in evaluating the 'sensitivity' of the building to noise. It is then necessary correctly to assess the levels of noise to be expected on a site and, in the case of noise coming from one direction, to plot these levels on the site survey. The periodicity and character of the noise, as well as actual level, should be noted.

It may be worth while listing some of the more important external noise sources to be considered

Aircraft: usually affecting all parts of a site equally and often from all directions; nuisance dependent on noise levels combined with periodicity; potentially the highest source of noise.

Road traffic: affects site differentially if from one direction; sound levels determined by traffic density and type, speed, road gradient, junctions and surrounding buildings; air-borne and ground vibrations.

Rail traffic: surface and underground; air-borne and ground vibrations; sound levels affected by embankments, cuttings and bridges; very variable in periodicity, frequency spectrum and sound level; invariably from one direction and affecting site differentially.

Factories: very variable in sound level and degree of annoyance; frequency spectra vary considerably; night shifts can seriously disturb residential areas.

Car parks: period of use affects seriousness of nuisance; 'revving up' and slamming of doors is main disturbance.

These and other sources of noise are very thoroughly evaluated in *Noise: Wilson Report* (H.M.S.O., 1963). The effect of rain on roofs of light construction is, however, generally overlooked in such surveys, yet can be very disturbing, particularly in the case of auditoria.

Having assessed levels of external noise, the next step will be to note against each item in the accommodation schedule

(1) acceptable intrusive noise levels for each room and
(2) levels of potential sources of noise within the proposed building.

These should again be marked on any plans produced (and the sections) so that the extent of any defence measures can be evaluated (see p. 115).

Noise sources within a building are so multifarious as to be practically impossible of enumeration. They may be broadly ascribed to the activities of the occupants and the operation of machines and equipment. Only a thorough knowledge of the working of a building can ensure that serious noise sources are not overlooked. In all cases the distinction should be made between air-borne noise and structure-borne vibrations.

Reduction of noise by planning

The accommodation schedule may now be broadly categorised under three headings

(1) rooms requiring quiet conditions, which we will refer to as 'quiet rooms'
(2) rooms containing a disturbing source of noise, referred to as 'noisy rooms and
(3) rooms, such as music practice rooms, which fall under both headings.

Using this rough classification, there are five kinds of defence measure which we can employ to reduce noise by planning.

(1) Quiet rooms can be located as far as possible from noisy rooms or from external sources of noise.
(2) The fenestration of quiet rooms may be arranged to face away from external sources of noise, or from noise emitted from noisy rooms.
(3) Quiet rooms may be defended from internal or external noise by rooms or circulation spaces in which noise is of little importance.
(4) Noisy rooms can be grouped so that their effect on other rooms is not so widespread.
(5) Structural transmission of noise may be interrupted by the physical separation of noisy rooms from quiet rooms.

It will be realised that most of the above measures operate in section as well as in plan. Reduction of noise by 'planning' is a three-dimensional matter, as will be seen from some of the examples which follow.

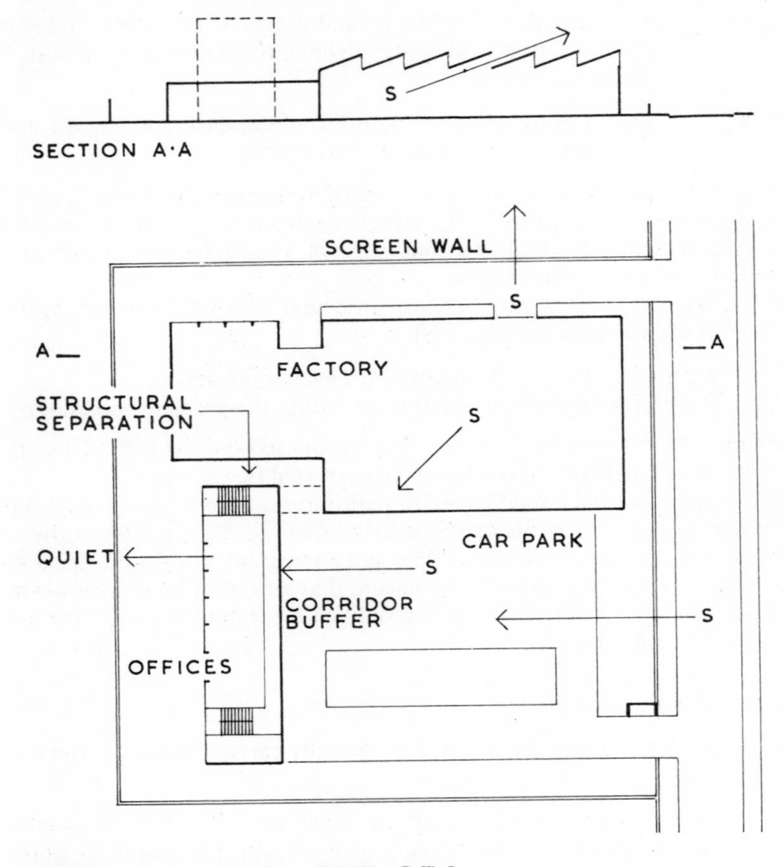

Figure 3.28

Figures 3.28 to 3.31 illustrate methods of reducing noise in the planning of four types of buildings, and draw attention to some of the external and internal noise sources which the architect may have to consider. The drawings are self explanatory and illustrate the use of the planning devices enumerated above. In a simplified form the drawings show the kind of analysis which, it is suggested, should be made in the early stages of all planning problems. The next step in the analysis would be to mark on the plans the estimated levels of noise in each case, together with the acceptable levels of noise for the various rooms, as shown in figure 3.36. The residual requirements of structural sound insulation will then be revealed. Potential sources of impact noise should also be marked on the sections.

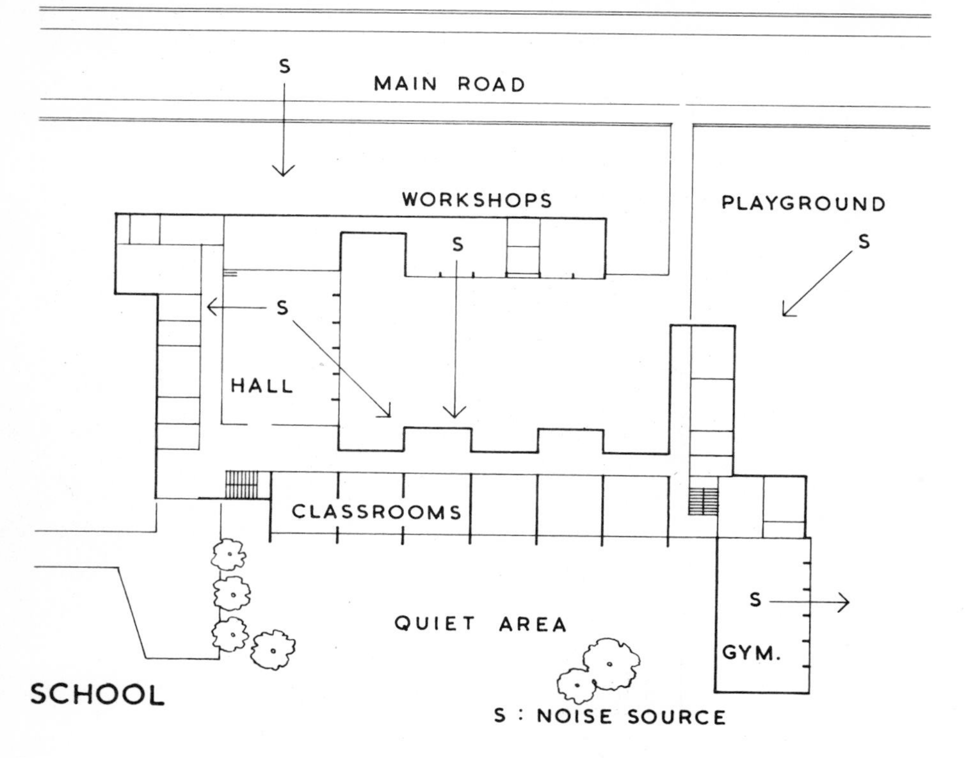

Figure 3.29

Town and regional planning

It is evident that noise is a major consideration, not only in the planning and construction of individual buildings, but also in the wider context of town and regional planning. The subject may be conveniently considered under the headings of regional planning, new town planning and urban renewal. The noise sources with which we will be concerned are, roughly in order of importance: road traffic and car parks, aircraft, airports and aircraft test beds, rail traffic and marshalling yards, hovercraft ports, factories and quarries, open-air swimming pools and children's playgrounds.

From a regional point of view we are firstly concerned with those arteries of communication which link urban and rural settlements and which in themselves generate noise, namely, trunk roads, air lanes and railways. In highly developed countries the patterns of trunk roads are well established and, in the main, run *through* towns and villages, causing increasing noise disturbance. Regional planning to remedy this is of necessity a long-term matter since, ideally, all trunk roads should, like motorways, run clear of built-up areas. It is, however, possible

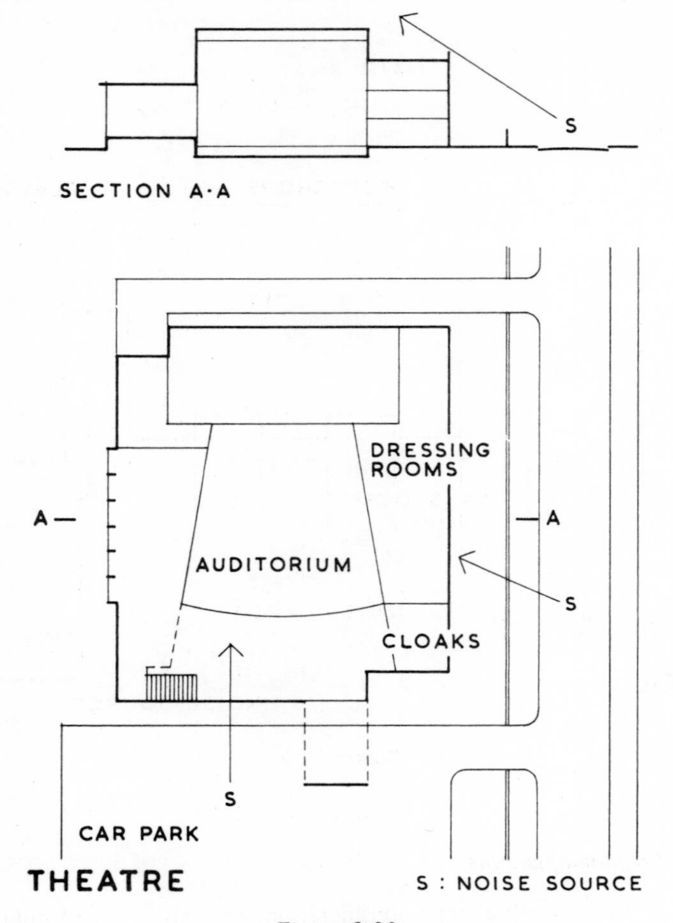

Figure 3.30

to site new towns well away from trunk roads and to see that existing towns and villages are not allowed to spread towards them. New trunk roads and motorways should, wherever possible, be sited well clear of existing settlements and existing trunk roads improved by the construction of by-passes.

It is now possible for planners to calculate, rather than guess at, the satisfactory distances required between trunk roads and built-up areas, taking into account the nature of the intervening terrain. In open, level country this distance is of the order of 300 m.

In advanced industrial countries operating many national and international air services it is becoming impractical to plan flight paths over settlement-free 'corridors'. High flying after leaving the vicinity of the airport is likely to be a more practical solution, regional planning taking into account the need to control building development near airports. New airfields would of course be sited in sparsely populated areas. In the long term, buildings vulnerable to noise near air-

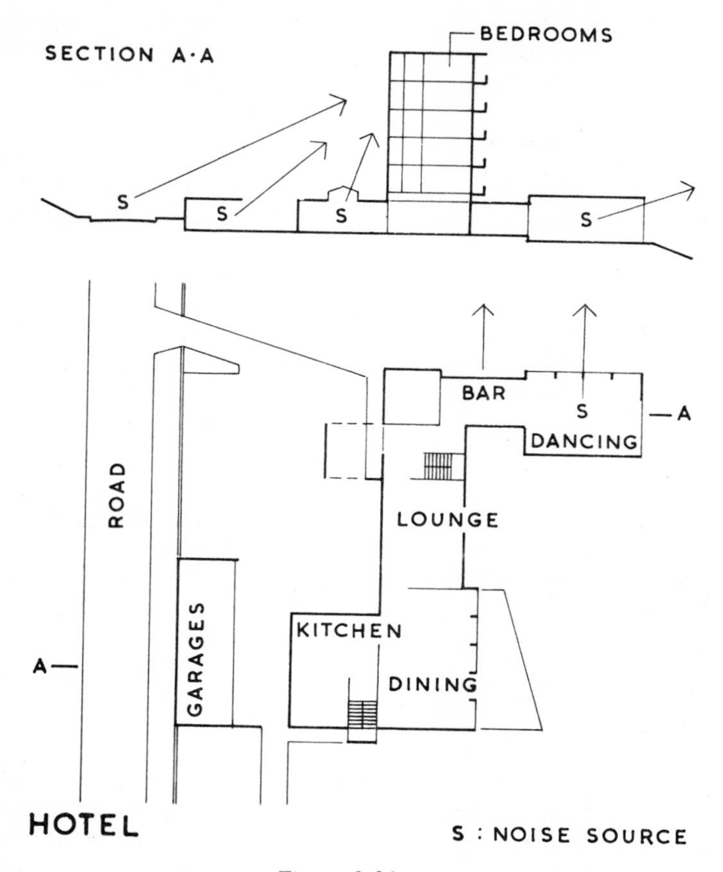

Figure 3.31

fields should not be replaced when they become obsolete.

Some idea of the extent of the area around a modern airfield where such building controls are necessary can be gained from a study of the noise disturbance contours around Heathrow Airport (see p. 105). Here the contour of serious noise disturbance (50 NNI) is 15 miles in extent in the direction of the main runways and about 3 miles in width.

As in the case of trunk roads, 'corridors' free of general building development should be provided for main railway lines and, in the case of existing railways, long term planning should have this as the ultimate aim. Marshalling yards are also a source of noise disturbance, particularly at night. In all cases of surface traffic it is, however, practical to protect near-by buildings from the worst effects of noise by screen walls (or a combination of cuttings and walls) provided only low-rise development is permitted. This may well provide a solution where railways penetrate built-up areas. Furthermore, with the electrification of railways it is now more practical to completely enclose sections of track as well as the stations.

The siting of industry so that noise disturbance is minimised should take into account two relevant factors: firstly the nature of the industrial process, and secondly the likelihood of night shifts being worked. The noise produced by many industrial processes can easily be contained within the factory buildings if this aspect is properly considered at the design stage. There is, from this point of view, no need to divorce all factories from other types of development. Where, however, an industrial process inevitably creates excessive noise in open sheds, yards and approach roads, such industries should be located well away from residential and other vulnerable buildings.

In this case no general guide can be given in the matter of distance from other types of building. Obviously, this will depend on the noise levels involved, and the criterion of acceptable noise level adopted. It is here that the possibility of night work has to be borne in mind since a much lower criterion will be necessary where the risk of sleep disturbance has to be considered.

New towns

Assuming that new towns are sited at a sufficient distance from trunk roads and airports, the main noise problem will be that arising from local traffic.

It is clear from recent surveys that average peak noise levels are determined, in the main, by

(1) mean traffic speed
(2) road gradients and junctions
(3) percentage of lorries and
(4) volume of traffic.

Traffic speeds are already limited in built-up areas and the choice of a reasonably level site will facilitate the reduction of road gradients. Road junctions also increase low-gear driving and this should be taken into account in the zoning of adjoining sites. The classification of lorry routes and the control of building development near them can reduce disturbance in those areas of the town where a high percentage of lorries is expected. In the pattern of roads suggested below, however, many lorries which at present pass through towns will find it unnecessary to do so.

The over-all volume of traffic in the new town will, however, be determined mainly by the size of its population so that the placing of a strict limit on the town's growth will be a major contribution to the reduction of noise.

In *Traffic in Towns* (1963), Colin Buchanan suggested a pattern of roads which, in effect, exploits the relationship between traffic density and noise level, with a view to reducing noise disturbance. He suggested that the urban road system should divide the town into major areas (which he called 'districts') and, within each district, a number of minor areas (which he called 'local areas'). The arrangement is shown diagrammatically in figure 3.32.

There would be three classes of roads: primary distributors, district distributors and local distributors. It will be seen from the drawing that primary distributors would be used by traffic entering the town and wishing to reach as quickly as possible its district destination, or by traffic moving from one district to another.

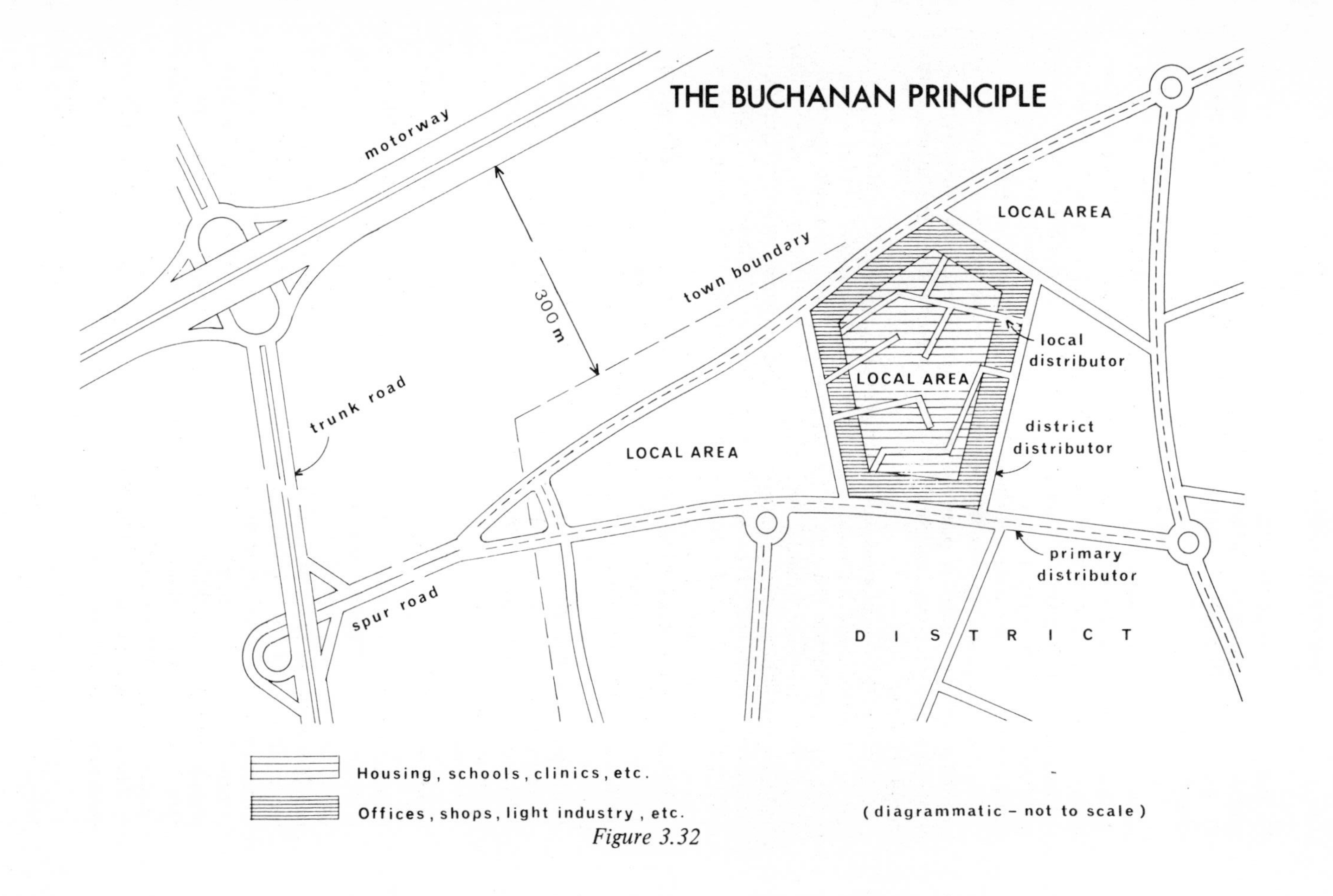
THE BUCHANAN PRINCIPLE
motorway
300 m
town boundary
trunk road
spur road
LOCAL AREA
LOCAL AREA
LOCAL AREA
local distributor
district distributor
primary distributor
D I S T R I C T
Housing, schools, clinics, etc.
Offices, shops, light industry, etc.
(diagrammatic – not to scale)

Figure 3.32

Having arrived at the desired district, traffic would use the district distributors to gain access to a local area. The local area would then be penetrated by means of the local distributors, all of which are cul-de-sacs to prevent their being used as district distributors.

That this arrangement shows many advantages will readily be seen but, in the context of the present subject, the important result is that, as traffic moves down the hierarchy of roads it also reduces in volume and therefore in average noise level. This imposes, as it were, a controlled pattern of traffic noise levels to which the zoning of building types can be related. Thus the most acoustically vulnerable buildings, such as housing, schools, clinics and libraries would be sited in the heart of the local areas where traffic density would be at its lowest.

Furthermore, buildings less affected by noise can be used to screen the heart of the local area by locating them along each side of the district distributors. Such buildings are: shops, car parks, light industrial buildings and offices with double glazing.

The majority of local areas would be residential with their local shops, primary school, clinic and branch library. The equivalent of local areas would, however, be allocated to heavier industry and groups of civic buildings, such as council offices, central library, hospital, secondary school and department stores. These buildings would be planned *facing inwards* around a quiet central precinct, with their service rooms and car parks on the periphery of the 'local area'. Generally speaking, the familiar road-facing design of buildings would be reversed and lay-out would be inward-looking.

Figure 3.32 also links the above concept with the major arteries of regional circulation, motorways and trunk roads. The motorways, as at present, are intended to serve long-distance traffic between regions, and the trunk roads provide access to, and between, towns. However, so that trunk road traffic would not pass *through* towns, as is fairly general at present, a sixth category of road is shown, the 'spur road'. The latter arrangement is fundamentally different from the by-pass method—which in effect creates an 'access loop' and is less discouraging to unnecessary entry.

Urban renewal

In this country, urban renewal has been largely ineffective in dealing with the high levels of road traffic noise now being generated. It was hoped that, by allowing traffic to flow more freely, by open planning and by building high, the noise problem would be reduced. In fact, the increased volume of traffic and the increased size and noise of lorries has out-weighed the advantage gained by a reduction in low-gear driving and the reduction of inter-reflected noise in narrow streets. The reduction in noise at the upper storeys of high buildings has been disappointingly small. Where road improvement schemes have not been put into effect, the increased volume of traffic has resulted in widespread penetration of residential areas by commuting traffic.

The proposals put forward by Buchanan, in part referred to above, may be summarised as follows.

(1) Rebuilding should follow a master-plan of the three categories of road: primary distributors, district distributors and local distributors.

(2) Primary distributors would be bordered by warehouses, light industries and other buildings least vulnerable to noise and separated from them by grass embankments and planting.

(3) District distributors would, in the main, be below pedestrian deck level and similarly bordered by car parks, service access to shops and storage spaces.

(4) All but essential local traffic would be excluded from the residential and shopping precincts, where also schools and hospitals would be sited.

(5) Pedestrian circulation would in most areas be raised to an upper deck, thus partly submerging local and service roads and screening buildings from noise.

The cost of such proposals and the time required to put them into effect as buildings become obsolete would be considerable. Meanwhile the rebuilding of large areas of our towns and cities has proceeded largely without regard to the proposals of the Buchanan report. However, since it is to be hoped that Buchanan's ideas may yet influence future development, it may be worth while to examine them from the point of view of their effectiveness in reducing noise disturbance.

There is no doubt that the screening of noise from main roads by buildings unaffected by noise can be very effective, provided the buildings to be protected are of similar height. The upper storeys of higher buildings would not necessarily benefit, however, and distance from the noise source may be insufficient. In all cases calculations of the screening effect are necessary, but this we are now in a position to do.

Submerged roadways present their own problems in the build-up of reverberant noise and the possibility of ground vibrations being transmitted to the foundations and superstructure of adjoining buildings. The use of sound-absorbing and vibration-damping materials may therefore prove necessary. Openings in pedestrian decks to provide natural light and ventilation for the roads below will become concentrated areas of escaping noise of which, as yet, we have little experience. The reverse arrangement, by which roads are carried at a higher level, can effectively screen pedestrians below from noise, but brings the source of noise nearer the upper levels of adjoining buildings. Balustrade walls along the edges of such roads will ameliorate the situation but their effect will depend on the relative position in section of wall and nearest windows.

While trees and shrubs do not form very effective screens to the transmission of air-borne noise, they can reduce sound reflections. Main roads bordered by trees, shrubs and grass embankments will be quieter for those using them because of the reduction of inter-reflected sound. This will also mean that the road, viewed as a horizontal sound source, will have a lower general noise level, to the benefit of adjoining buildings.

When all such measures have been adopted, there will no doubt remain many buildings bordering traffic routes which will require heavy walls and double-glazed windows. The advantage of the Buchanan proposals is that such buildings are kept to a minimum. However, up to the present, the most effective form of noise control is that operated by local authorities in refusing to grant planning permission for housing unless steps have been taken to reduce internal noise levels to a *maximum* of 50 dB.A, either by layout or double glazing.

The problem of noise in existing and new towns can, however, be tackled in another way which, in the writer's opinion, would be more practical, less expensive and more effective. This would be to drastically reduce traffic entering towns and to reduce the noise emitted by such vehicles as are necessary. Peripheral car parks together with quiet, ample and free public transport would reduce the volume of traffic; submerged service roads could separate lorries and delivery vans from public and private transport. Almost silent modes of transport are now technically practical and, although such planning proposals as have been outlined above will still be valuable in terms of road safety, the noise problem could be solved by its reduction at the source.

Measurement and Calculation

The remainder of this section will be devoted to typical calculations in the field of noise control. Calculations of the effectiveness of any measures taken to reduce noise disturbance will be concerned with:

(1) the source of sound
(2) the paths of transmission
(3) the requirements of the listener.

In practice, the source of sound is usually a variable quantity, the paths of transmission are often complex, and the requirements of the listener are difficult to determine. For these and other reasons calculations can only be more or less approximate and must be based, in part, on value judgements regarding acceptable standards.

Specialists in the field of applied acoustics are continually seeking more reliable techniques of measurement and calculation. There is therefore a tendency to proliferation of methods of calculation which is confusing to the non-specialist. A limited selection has therefore been made of those calculations likely to be most used by architects and town planners.

Instruments for the measurement of noise

For the purpose of design, the architect will either refer to published measurements of typical noise sources, such as are given on p. 108, or will himself take such measurements with suitable instruments. A brief description of three instruments used in the measurement of noise is in any case necessary to an understanding of the calculations that follow.

The first of these is the 'sound level meter', a portable instrument for measuring over-all sound intensity levels. Essentially, this consists of a built-in microphone, an amplifier, and an indicating meter from which sound levels can be read in decibels. Incorporated in the amplifying system may be 'weighting networks' which can be switched in if required. The purpose of these is to modify the amplification of the frequency components of the sound measured to correspond approximately to the varying sensitivity of the ear at different frequencies. Since this variation of sensitivity changes as sounds become louder, three weighting networks (A, B and C) are incorporated, the A network being switched in for sounds below 55 dB, the B network for sounds between 55 and 85 dB, and the C network for sounds above 85 dB. The reduction in response, with frequency, is shown in figure 2.4 on p. 43 in respect of the A network.

For pure tones, using the appropriate weighting network, readings on the meter will therefore correspond fairly closley to phons. For complex sounds this correspondence is more approximate. Nevertheless, for the comparison of two similar noises (such as traffic noise at two different situations) the meter provides a good guide to relative loudness. The limitations of the instrument become evident when, for example, aeroplane noise is compared with road traffic noise.

When the sound level meter is used without the introduction of weighting networks, its response is said to be 'flat' and the reading is stated simply in decibels. When a weighting network is employed, this should be stated, for example, 45 dB.A or 60 dB.B. In all cases the distance between the noise source and the microphone must be recorded. Noise levels in rooms will normally be a measure of the reverberant sound level, averaged for different positions in the room, and stated as, for example, 85 dB.C (RSL).

Because of the need to simplify calculations (and incidentally the taking of measurements) it has in recent years become general practice to use the 'A' weighting network for road traffic noise measurements, regardless of over-all noise level. It has been found that, in respect of this type of noise disturbance, the correlation between dB.A levels and subjective loudness is adequate for design purposes. For this reason many of the sound level meters now produced for simple field work give readings only in dB.A.

Nevertheless, the use of more accurate measurements of noise sources is often necessary and is employed in some of the calculations that follow. For these it is necessary to know the frequency spectrum of the noise. A more elaborate, but still portable, instrument is available, called an 'octave band sound level meter', which will provide a frequency spectrum for use in 'six-figure calculations'. The instrument fulfils all the functions of the sound level meter first described above but, in addition, is provided with filters by which the sound intensity level in any one of a number of octave bands can be measured. As each filter is switched in, the sound level is indicated (in dB) by the meter for that particular octave band of frequencies.

One such meter is provided with six filters for octave bands centred on the following frequencies

125 250 500 1000 2000 4000 Hz

together with 'high pass' and 'low pass' filters for frequencies above and below these octave bands. The result of such an analysis can conveniently be plotted on a graph, such as the one given in figure 1.9 to show the distribution of sound energy in terms of frequency.

A third type of instrument, one which records fluctuating sound levels over an extended period of time, is described below.

Fluctuating noise levels

Most noise sources fluctuate in intensity, ranging from the fairly steady noise of an electric motor to the widely varying noise levels of road traffic. Such variations present a difficulty in design since the primary element in any calculation of sound insulation—the intensity of the sound source—is not a fixed quantity.

Quite obviously a decision has therefore to be made whether, for design purposes, to work from peak noise level, average noise level, or some other evaluation of the noise source. This decision will be influenced by the circumstances of each case. For example, in insulating a recording studio from aircraft noise, it would be sensible to do this in relation to peak noise levels. On the other hand, in considering the interference of traffic noise with the comfort of office workers, it would seem hardly necessary to take account of the very occasional peak noise of a high-powered motor cycle.

Road traffic noise

Many studies have been made of the relationship between people's response to noise and the character and degree of fluctuation of various types of source. Response may be measured in terms of apparent loudness or degree of disturbance and will of course vary from individual to individual. Such studies have produced the concepts of 'noise climate', 'traffic noise index' and the '10% level'—in respect of road traffic noise. Since the latter concept is at present more generally used and has been adopted by the Government for the purpose of building development approvals, it will be described here and used in subsequent calculations.

The degree of disturbance caused by road traffic depends mainly on its loudness and its variability with time. If measurements are made in dB.A assessment of loudness will have been made in a situation where the frequency spectra of vehicles vary. Studies have also shown that there is a good correlation between degree of disturbance and the dB.A level exceeded for 10% of the time—exceeded, that is, by the peak noise levels. The period over which noise levels are measured is 18 h, between 06.00 and 24.00 hours on a typical weekday, although measurements taken over shorter periods can provide a good approximation to what has become known as the L_{10} (18 h) index.

An instrument is now available which, after being switched on, will record the fluctuating noise levels of passing vehicles and at the end of the period of operation provide the L_{10} figure for that period.

A typical example of the L_{10} noise level would be about 75 dB.A at 10 m from the edge of a busy main road through a residential area. In this situation peak noise levels would be 75 dB.A *and above* for 10% of the time. Such a level would be used for contouring a site (see p. 88) and subsequent calculation of sound insulation for its buildings.

Aircraft noise

The frequency spectrum of mixed road traffic noise presents a fairly smooth curve when plotted on a graph, albeit rising at frequencies around 100 to 500 Hz (see p. 108). Aircraft noise on the other hand has components which predominate to a marked degree in particular frequency bands. These frequency components can have a considerable effect on subjective impressions of loudness, for which the dB.A scale would not adequately allow.

Measurements of aircraft noise are therefore made, in dB, in each of a number of narrow frequency bands and from these a calculation of loudness is made which gives due emphasis to the dominant frequency components. This level is referred to as the 'perceived noise level' and expressed as PNdB. The relationship between PNdB and dB.A varies with the varying spectra of aircraft noise but PNdB values are roughly 13 units higher than dB.A levels.

Disturbance caused by aircraft noise varies, as would be expected, with loudness and the number of flights heard in a given period of time. It may also be increased by the tensions built up in the listener as aeroplanes approach and by other psychological factors.

By relating people's reactions to aircraft noise, at various positions near Heathrow Airport, to the measured levels and periodicity of overflights in those positions, a relationship was found, as expressed by the following formula, in which 'degree of annoyance' is expressed as a 'noise and number index' (NNI)

$$\text{NNI} = \text{av. peak PNdB} + 15 \log_{10} N - 80$$

where average peak noise level is the logarithmic average and N is the number of flights heard between 06.00 and 18.00 hours in summer time.

The relationship between the NNI indices and people's reactions during the daytime is as follows

NNI 65 = unbearable
50 = very annoying
35 = annoying
25 = intrusive
10 = noticeable
0 = not noticeable

The results of the Heathrow survey (1961)* were plotted as contours on a map of the area. On the axis of the main runways the 50 NNI contour extended some 4 miles on each side of the centre of the airfield and embraced the built-up areas of Hounslow and Horton.

Using the formula, NNI contours can be drawn without repeating the social survey, if the known or expected noise levels and flight paths are available. This was done for the projected third London Airport in order to evaluate relative disturbance of built-up areas on the four sites considered. In the same way contours are up-dated for Heathrow Airport and in 1972 the 50 NNI contour extended some 7 miles on each side of the centre of the airfield.

Figure 3.33 shows a hypothetical pattern of NNI contours around an imaginary airport, in order to illustrate the effect of runway positions and flight paths on the shape of the contours. The extended nature of the contours are caused by the predominant use of the E–W runway. The 'bulges' in the contours are caused by the less frequent use of the NE–SW runway and the various flight paths on leaving the vicinity of the airfield. The general scale of the contours is realistic for a busy airport.

NNI contours can have a number of uses

(1) for the comparison of alternative sites for new airfields
(2) for planning control in the vicinity of existing airfields

**Noise: Wilson Report* (H.M.S.O., 1963)

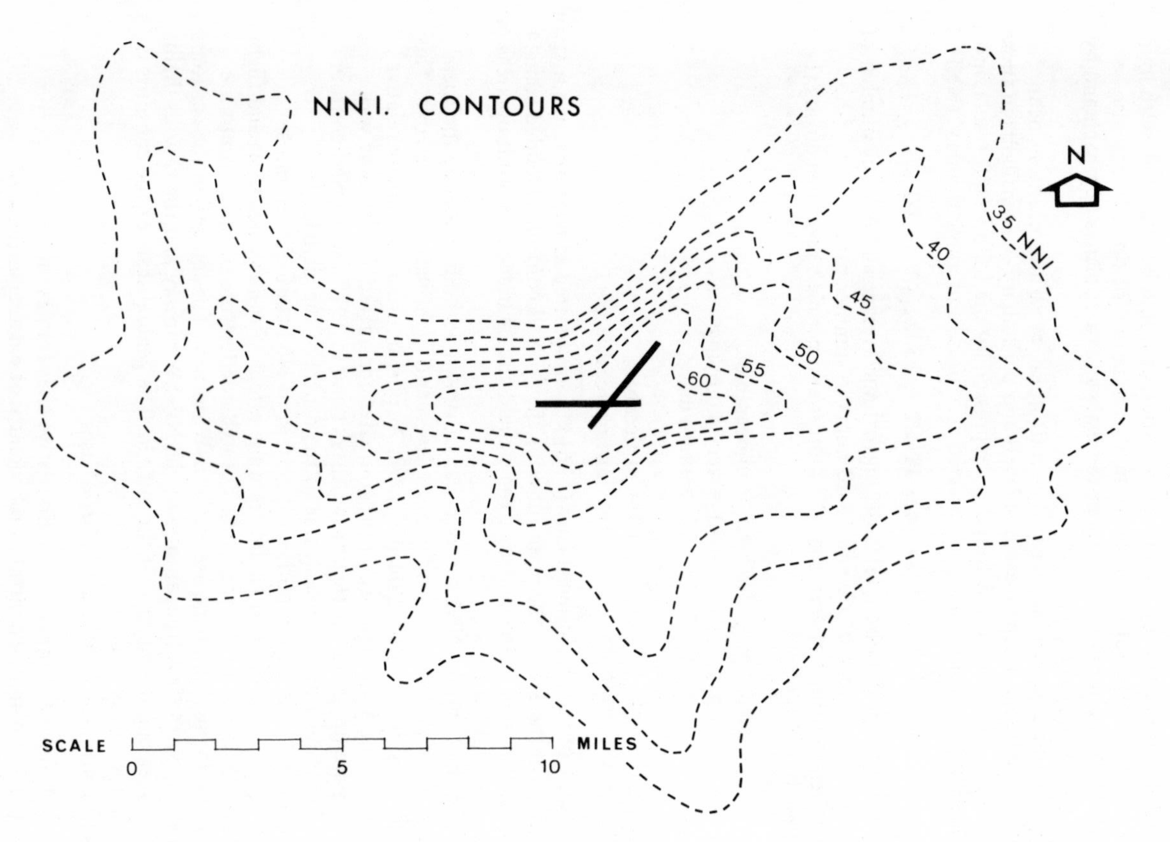

Figure 3.33

(3) for the specification by local authorities of standards of sound insulation for new buildings near airfields and

(4) for the making of Government grants towards the sound insulation of existing buildings.

Industrial noise

Industrial noise is of course very varied in over-all intensity, frequency spectra and periodicity. In some processes sound levels well over 100 dB can be reached, as measured within the factory. Frequency spectra vary widely and the noise can contain concentrations of energy in very limited frequency bands and sometimes strong pure tones. The noise can be steady and continuous, it can fluctuate in intensity or frequency, it can be impulsive with a regular or irregular rhythm or it can occur for periods and at intervals of any duration.

Where the noise is steady and continuous it can be measured in dB.A and compared with the criteria of acceptable intrusive noise such as are listed on p. 112. This form of measurement will not, however, provide an assessment of noise disturbance in the case of noise with some of the other characteristics listed above. In such cases, degree of disturbance is normally assessed by adjusting the dB.A figure (up or down) to take into account the *character* of the noise and other relevant factors. Alternatively, and more accurately, a frequency analysis of the noise can be made, providing a 'noise rating' which itself can be adjusted in a similar fashion. If the former method is employed, the adjusted dB.A value is referred to as the 'corrected noise level' and in the latter case the term 'corrected noise rating' is employed. The various adjustments are described on pp. 134 to 135 in association with example calculations.

Frequency spectra

As has been explained on p. 67 under 'average insulation values', a more reliable calculation of insulation requirements can be made if the frequency spectrum of the offending noise can be obtained. Some idea of the variation in frequency spectra can be seen from the list of examples in table 3.1, for which an octave band analysis has been made. It should, however, be stressed that such an analysis does not reveal possible concentrations of energy in narrower frequency bands, or the presence of strong pure tones.

Addition of two or more noises

When two pure tones of equal intensity are produced at the same time, the sound intensity level of the combined sound will be 6 dB higher than either of them, if they are of the same frequency and if the sound waves are in phase. If on the other hand two 'noises' of equal intensity level occur simultaneously, the resulting sound will be 3 dB higher than either one of them, provided the noises have a wide and random frequency pattern. If the two noises are unequal in sound level

TABLE 3.1 Octave Band Analysis of Some Typical Noises (dB)

Mid-frequencies of octave bands (Hz)	125	250	500	1000	2000	4000	*Total*
Accounting machines in reverberant office	72	74	76	77	77	76	**83**
Aircraft, four-engine jet 40 m overhead	121	123	124	122	119	116	**129**
Diesel generator, 1700 hp in reverberant room	82	88	82	83	79	75	**92**
Electric train, in the open, 3 m distance	97	96	94	92	85	78	**101**
Metal saw, machine saw in reverberant workshop	75	84	86	93	105	106	**109**
Pneumatic drill without muffler, at 3 m distance	94	93	89	91	92	88	**99**
Printing works, reverberant workshop	90	91	90	91	90	84	**97**
Riveting, large steel plate, at 3 m distance	94	101	103	107	106	110	**114**
Road traffic, heavy urban, at 10 m distance	77	76	73	67	62	58	**81**
Telephone bell at 3 m distance	38	49	58	64	72	62	**73**
Typing pool, ten typists, reverberant room	65	63	61	60	60	56	**70**
Weaving shed, reverberant workshop	92	95	95	96	98	97	**104**
Wood planer, planing machine in reverberant workshop	88	94	101	102	102	95	**107**
Wood saw, circular saw in reverberant workshop	71	72	78	78	86	88	**91**
Works canteen, plaster ceiling, average RSL	55	61	67	65	59	54	**70**

the resulting sound will have a level *less* than 3 dB in excess of the higher one, in accordance with the scale shown in figure 3.34.

Thus, if for example road traffic noise of 71 dB is heard together with the background noise in a canteen of 65 dB, the resulting noise level will be 72 dB.

This method of calculation can also be used to obtain the approximate overall sound level from a frequency analysis of the sound. Table 3.2 gives the frequency spectrum of a noise in octave bands and, below it, the addition of the values by adding together the first pair, then adding this to the third value, and so on. It will be seen that the final total is 3.4 dB above the highest octave band value. Having in mind that sound energy below and above the six octave bands has not been included, we may round the figure of 71.4 up to 72 dB.

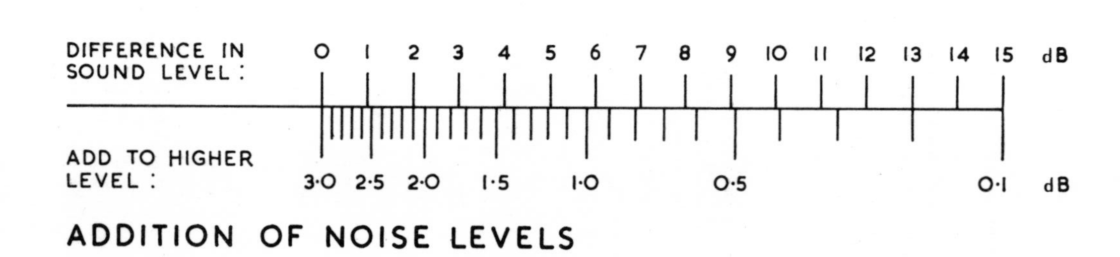
DIFFERENCE IN SOUND LEVEL :
0 1 2 3 4 5 6 7 8 9 10 11 12 13 14 15 dB
ADD TO HIGHER LEVEL :
3·0 2·5 2·0 1·5 1·0 0·5 0·1 dB
ADDITION OF NOISE LEVELS

Figure 3.34

TABLE 3.2

Mid-frequencies of octave bands (Hz)	125	250	500	1000	2000	4000
Sound intensity levels (dB)	55	60	68	67	60	54

60.0 – 55.0 = 5.0 therefore add to higher level 1.2 = 61.2 dB
68.0 – 61.2 = 6.8 therefore add to higher level 0.8 = 68.8 dB
68.8 – 67.0 = 1.8 therefore add to higher level 2.2 = 71.0 dB
71.0 – 60.0 = 11.8 therefore add to higher level 0.3 = 71.3 dB
71.3 – 54.0 = 17.3 therefore add to higher level 0.1 = 71.4 dB

Conversion from dB to dB.A

If the frequency spectrum of a noise is known, it is possible to convert these values into the equivalent of a dB.A reading on a sound level meter. Reference to figure 2.4 will show that the values in each octave band have to be adjusted as shown in table 3.3. We have taken our previous noise spectrum as an example so that we can compare the result in dB with that in dB.A. We find that, whereas the over-all sound intensity level was 72 dB, expressed as a loudness level it is 70 dB.A.

TABLE 3.3

Mid-frequencies of octave bands (Hz)	125	250	500	1000	2000	4000
Sound intensity levels (dB)	55	60	68	67	60	54
A weighting correction (dB)	–16	–9	–3	0	+1	+1
Corrected values	39	51	65	67	61	55

51.0 – 39.0 = 22.0 therefore add to higher level 0 = 51.0 dB
65.0 – 51.0 = 14.0 therefore add to higher level 0.2 = 65.2 dB
67.0 – 65.2 = 1.8 therefore add to higher level 2.1 = 69.1 dB
69.1 – 61.0 = 8.1 therefore add to higher level 0.6 = 69.7 dB
69.7 – 55.0 = 14.7 therefore add to higher level 0.1 = 69.8 dB

Structural sound insulation values

The simplest calculation in noise control is where, having established the level of noise at the outside face of a building, a form of construction is adopted which will reduce the noise to an acceptable level for the occupants. Table 3.4 provides the necessary information in respect of typical building elements. Since insulation varies with frequency the values are given at six frequencies and these are

TABLE 3.4 Air-borne Sound Insulation Values

Walls and partitions (Hz)	125	250	500	1000	2000	4000	av.
Brick 340 mm, plaster one side	44	46	50	57	61	63	**54**
Brick 230 mm, plaster one side	41	45	48	56	58	61	**52**
Brick 115 mm, plaster both sides	34	36	41	51	57	60	**47**
Concrete 150 mm, dense	36	39	45	52	60	65	**50**
Clinker block 75 mm, plastered	26	33	40	48	56	58	**44**
Clinker block 50 mm, plastered	25	29	33	38	47	57	**38**
Clinker block, 2 leaves each 75 mm (50 mm air space outer faces plastered)	41	44	47	52	61	69	**52**
Clinker block, 2 leaves each 50 mm (50 mm air space outer faces plastered)	40	44	45	49	57	65	**50**
Wood studs 100 × 50 mm at 400 mm c/c, plastered wood laths both sides	35	24	34	37	45	61	**39**
Wood studs as above but 13 mm dense fibreboard both sides	16	22	28	38	50	52	**34**
Floors							
Wood joists, t and g boarding, plaster board ceiling, skim coat plaster	18	25	37	39	45	45	**35**
Wood joists as above but boards on battens on fibreglass quilt	25	33	38	45	56	61	**43**
Wood joists as above but with 50 mm sand on 3-coat plaster and metal lath	36	42	47	52	60	63	**50**
Concrete 130 mm and 25 mm screed and plaster soffit	37	38	43	51	60	64	**49**
Concrete 130 mm and 65 mm floating screed on fibreglass quilt and plaster soffit	39	44	49	55	62	64	**52**
Concrete and hollow tile 150 mm and wood floor on 50 × 50 mm battens and plaster soffit	36	38	39	47	54	55	**45**
Doors							
Wood 50 mm solid normal hanging	12	15	20	22	16	24	**18**
Wood 50 mm solid airtight	15	18	21	26	25	28	**22**
Windows							
Single 4 mm glass, tightly closed	17	21	25	26	23	27	**23**
Single 6 mm glass, tightly closed	20	24	28	29	26	30	**26**
Double 4 mm and 6 mm glass tightly closed and 180 mm air space with absorbent linings	23	37	41	34	39	35	**35**
Double 4 mm glass fixed, 200 mm air space with absorbent linings	30	35	43	46	47	39	**40**
Double 9 mm glass fixed, 340 mm air space with absorbent linings	31	38	43	49	53	63	**46**

averaged in the last column, for use in single-figure calculations.

The insulation values are for the stated structural elements themselves and do not take into account loss of insulation due to flanking transmission (see p. 76). This will be negligible if the flanking structure is of comparable weight. The figures are also relevant to a reverberation time of 0.5 s in the receiving room and to structures of domestic or office dimensions.

Acceptable intrusive noise levels

The third value which needs to be established in any sound insulation calculation is of course the criterion of the level of noise which is acceptable in any given situation, at least to the majority of people. This is called the 'acceptable intrusive noise level' and its meaning and some of the factors which influence it have been discussed in section 2, p. 45.

Such criteria can be expressed either in dB.A (for single-figure calculations) or as a 'noise rating', NR (for six-figure calculations). In the latter case reference is made to 'noise rating curves', as in figure 3.35, which provide a range of intrusive noise 'ceilings' over six frequency bands, appropriate to various criteria. Having decided on the appropriate criterion, for example NR 45, then the intrusive noise should not rise above the levels indicated by the NR 45 curve. It will be seen that criteria in this form can be related to the frequency spectrum of a noise and also the insulation values of a structure at six frequencies.

Table 3.5 gives suggested criteria in both dB.A and NR for a variety of situations, based on the results of surveys by a number of authorities, whose results show a fair measure of agreement. The range of values is of course influenced mainly by the normal ambient noise level in the receiving room and the use to which the room is put. The criteria are ideal and relate to meaningless and fairly continuous noise such as that from road traffic.

TABLE 3.5 Acceptable Intrusive Noise Levels

	dB.A	NR		dB.A	NR
Banks	50	40	Libraries, loan	45	35
Churches	35	25	Libraries, reference	40	30
Cinemas	35	25	Music rooms	30	20
Classrooms	35	25	Offices, private	40	30
Concert halls	30	20	Offices, public	50	40
Conference rooms	30	20	Open-air theatres	40	30
Court rooms	35	25	Radio studios	30	20
Council chambers	35	25	Restaurants	50	40
Department stores	55	50	Recording studios	30	20
Flats, living	45	35	Shops	55	50
Flats, sleeping	35	25	Telephoning, good	50	40
Hospitals, wards	35	25	Telephoning, fair	55	45
Hotels, bedrooms	35	25	Television studios	35	25
Houses, living	45	35	Theatres	30	20
Houses, sleeping	35	25	Typing pools	55	50
Lecture rooms	35	25	Works canteens	60	55

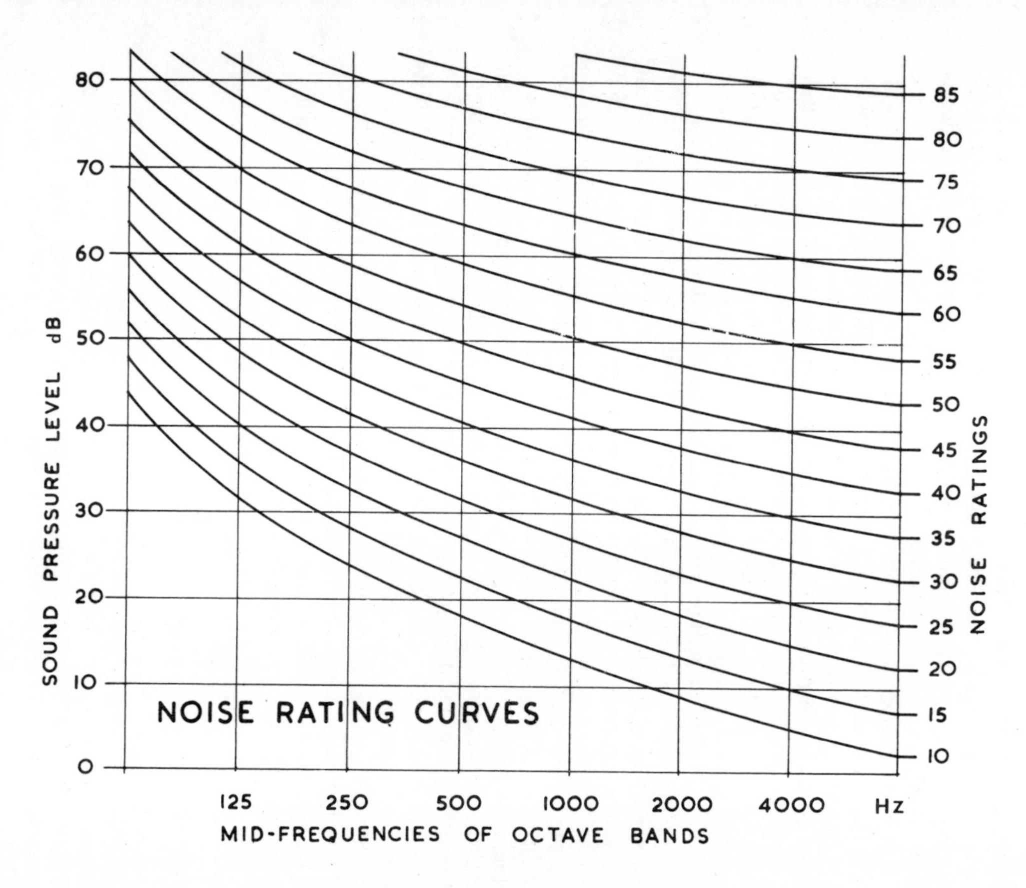
NOISE RATING CURVES
SOUND PRESSURE LEVEL dB
80
70
60
50
40
30
20
10
0
125
250
500
1000
2000
4000
Hz
MID-FREQUENCIES OF OCTAVE BANDS
NOISE RATINGS
85
80
75
70
65
60
55
50
45
40
35
30
25
20
15
10

Figure 3.35

Single-figure Calculations

Road traffic noise

Figure 3.36 shows a school site on which L_{10} traffic noise contours have been plotted, taking into account reductions due to distance only. A preliminary plan for the school is shown for the purpose of evaluating approximately the degree of disturbance in rooms exposed to traffic noise.

The building itself will of course modify the noise levels on the site, as will other factors described below. For a first evaluation of the noise problem we may, however, assume that, for example, the noise external to the school hall will be about 69 dB.A and, external to the classroom windows, about 64 dB.A.

We now mark on the plan the optimum acceptable intrusive noise levels of 30 dB.A for the hall and 35 dB.A for the classrooms. By deduction, we then find that the external wall of the hall should have an average insulation value of 69 – 30 = 39 dB. This will present no difficulty if the wall is of traditional construction and is without windows (see p. 111). If windows are required then they should be of fixed double glazing to provide an insulation of about 40 dB.

The insulation required for the classrooms is approximately 64 – 35 = 29 dB. As will be seen from the more detailed calculation on p. 125, the required insulation can be achieved by 6 mm *fixed* glazing set in a wall of traditional construction if window to wall ratio is 2:1. This does, however, mean that an alternative method of ventilation will be needed. This first approximate calculation therefore exposes the economic problem of providing an alternative method of ventilation for most of the rooms on this side of the building.

The plan also indicates a few of the estimated internal noise levels and, if we take the classrooms as an example, we see that the partitions between them should have a minimum insulation value of 75 – 35 = 40 dB. This could be achieved by using, say, 75 mm clinker blocks plastered on both sides.

We now have to consider some of the factors, other than distance, which would modify the noise levels shown by the contours in figure 3.36.

Angle of view

The source of noise in the case of heavy road traffic approximates to a 'line source'. The classrooms shown in our preliminary plan for the school will be partially shielded from this extended source by the projecting wing of the assembly hall block, and the classrooms nearest this wing will benefit most. Table 3.6 gives the reductions to be made to allow for this partial shielding, in terms of what is called 'angle of view'. This is the angle subtended by the length of road which is visible from the 'reception point'–in this case, from any given classroom. Assuming no building on the adjoining site to the east, the classroom marked as having an external noise level of 64 dB.A will have an angle of view of 140° and therefore a reduction of 1 dB.A should be made.

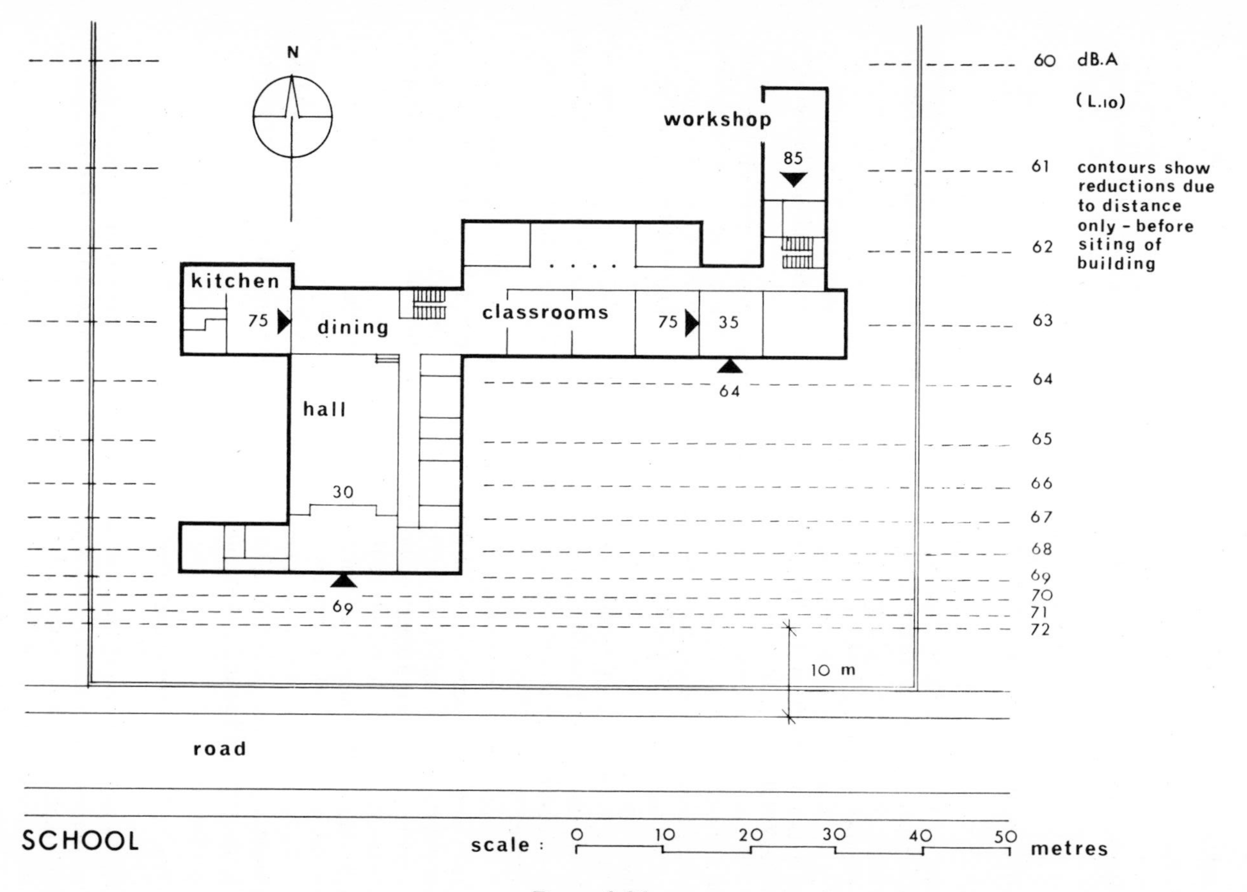
N
workshop
85
kitchen
75
dining
classrooms
75
35
64
hall
30
69
60 dB.A
(L.10)
61
contours show reductions due to distance only - before siting of building
62
63
64
65
66
67
68
69
70
71
72
10 m
road
SCHOOL
scale : 0 10 20 30 40 50 metres

Figure 3.36

TABLE 3.6*

Angle of view	180°	140°	110°	90°	70°	56°	44°	35°	28°	22°	18°
Reduction (dB.A)	0	1	2	3	4	5	6	7	8	9	10

*Data from *Calculation of Road Traffic Noise* (Department of the Environment, 1975).

Effect of paving, grass and foliage

In contouring the school site, no allowance was made for the fact that, in relation to ground floor accommodation, the path of sound would be near the ground and therefore affected by it. If the space between road and classrooms is mainly paved, as for a playground, then an increase of about 2 dB.A can be expected, as compared with sound propagation in a 'free field'. This increase is not however cumulative with distance; it is a 'once for all' increase due to reflection from the hard surface and may already have been included in meter readings taken near the road.

If, however, the space between road and building is mainly grassed, there will be a cumulative absorption of sound at a rate dependent on the mean height of the path of propagation between source and reception point. Figure 3.37 shows the allowances that can be made for what is called 'ground absorption' in the case of grass.

It will be seen that, in the case of the ground floor classrooms considered above, the figure of 64 dB.A should be reduced by about 4 dB.A if the intervening area between road and building is mainly grass and if the mean height of sound propagation is 0.75 m.

The planting of trees between the source of sound and a building causes a much smaller reduction of noise than would be expected. Where the path of propagation is through dense foliage a reduction of about 1 dB.A per 10 m can be expected, so that a single row of trees will have little effect on the level of noise. Near ground level, where the sound is passing below the tree foliage, the reduction will be negligible. Furthermore, if the trees shed their leaves in winter sound reduction will be even less than that given above.

To be effective, therefore, a belt of closely spaced, evergreen trees should be planted to a depth of at least 30 m and such planting should be associated with shrubs to absorb sound near ground level.

Effect of air absorption and wind

The reduction in sound level due to air-absorption varies with frequency, temperature and humidity but is very small for distances with which the architect is normally concerned. For example, at 20 °C and 60% relative humidity absorption will be less than 1 dB.A over a distance of 100 m.

The effect of wind gradients on the propagation of sound is a complicated phenomenon, affected by associated air turbulence and temperature gradients. Changes in sound level are not simply related to distance and vary

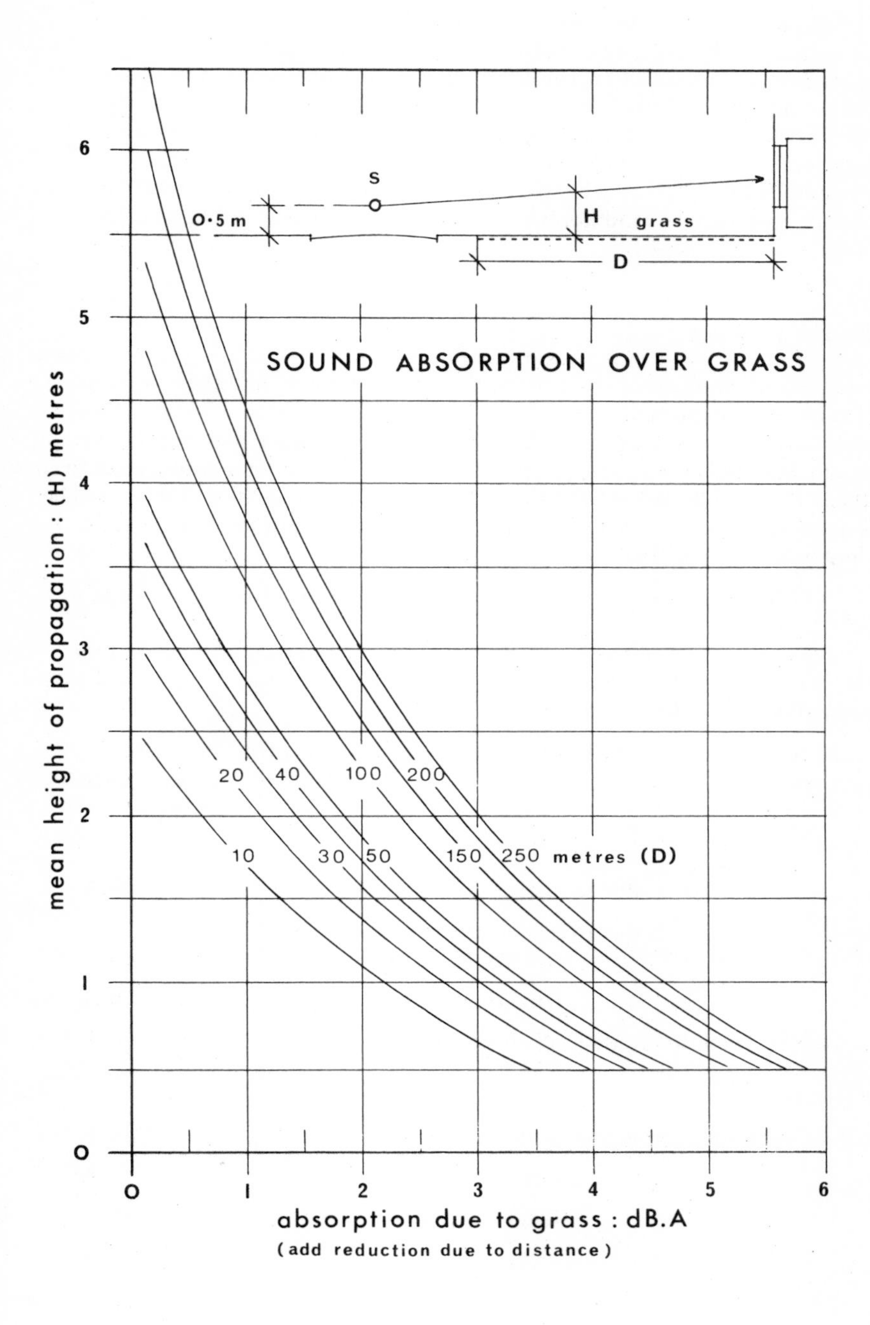

SOUND ABSORPTION OVER GRASS
S
0·5 m
H
grass
D
10
20
30
40
50
100
150
200
250
metres (D)
mean height of propagation : (H) metres
absorption due to grass : dB.A
(add reduction due to distance)
0
1
2
3
4
5
6

Figure 3.37

in a complex manner with frequency. For distances up to about 150 m fluctuations of the order of 0.8 dB.A per 10 m of travel can be expected for winds of 10 miles/h. In other words, down-wind over a distance of, say, 50 m sound may be 4 dB.A higher than if the air is at rest. Up-wind it may be 4 dB.A less. A wind at right angles to the path of sound will have no appreciable effect. Over a period of time therefore the average effect of changes in wind direction is likely to be small and is not normally taken into account, especially as wind noise is likely to have a masking effect on other noises.

Effect of screening

The effect of an obstruction between the source of sound and the reception point has been explained on p. 89 and by figure 3.26. Figure 3.38 gives the reductions to be expected in dB.A relative to the difference between the diffracted path of sound $(x + y)$ and the direct path (z).* A small reduction occurs even outside the 'sound shadow' and this is shown to the left of SB on the scale. The sectional drawing is an example to show the relevant positions of x, y, and z. It is important to note that

(1) the reductions given by the scale are additional to reduction due to distance, and

(2) the reductions apply to screens extended laterally well beyond the reception point in the case of road traffic, or otherwise designed to cut off sound passing the end of the screen.

The section in figure 3.38 can be used as an example of the calculation. Assuming the L_{10} level of traffic noise to be 75 dB.A at 10 m distance, then

At ground floor

Reduction due to distance of 19.90 m	= 4 dB.A
Reduction due to screen: $(x + y) - z$ = $(8.25 + 11.65) - 19.65 = 0.25$ m	= 11 dB.A
Total reduction	= 15 dB.A
Therefore noise level at window C	= 60 dB.A

At first floor

Reduction due to distance of 21.10 m	= 4 dB.A
Reduction due to screen: $(x + y) - z = 0$ m	= 4 dB.A
Total reduction	= 8 dB.A
Therefore noise level at window B	= 67 dB.A

*Data from *Calculation of Road Traffic Noise* (DOE, 1975)

SOUND REDUCTION DUE TO SCREENING

(add reductión due to distance)

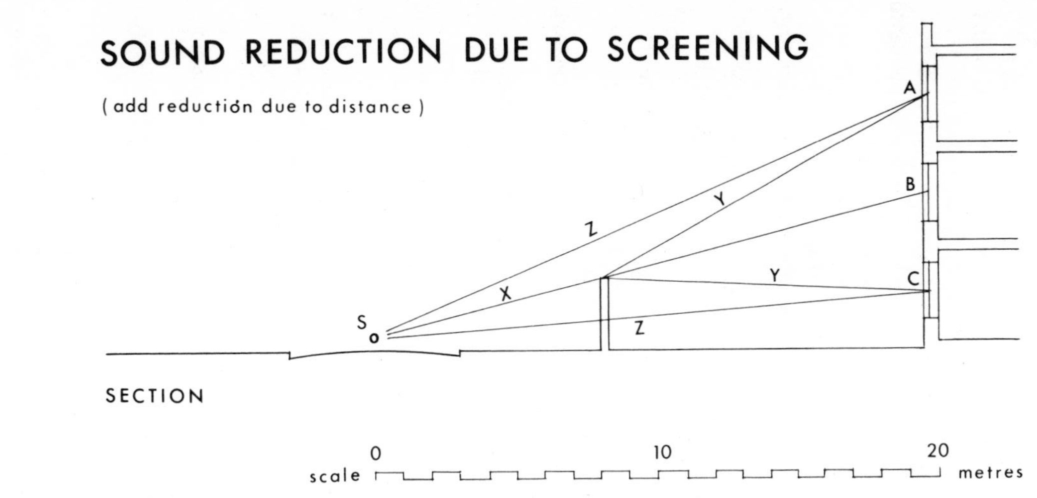

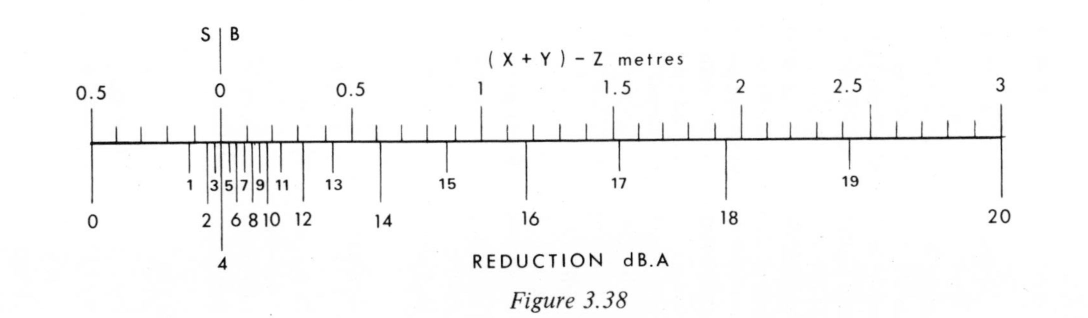

Figure 3.38

At second floor

Reduction due to distance of 21.25 m	=	4 dB.A
Reduction due to screen: $(x + y) - z$ $= (8.25 + 13.10) - 21.25 = 0.1$ m	=	1 dB.A
Total reduction	=	5 dB.A
Therefore noise level at window A	=	70 dB.A

Figure 3.39 shows the ground floor plan of the building and screen wall in figure 3.38. The screen wall is terminated at D to illustrate the effect of this on noise levels at windows E, F, G and H. By calculation* it can be found that the noise reduction due to screening is reduced to 20% of the screen's potential at window E, 30% at window F, 45% at window G and 55% at window H (disregarding sound reaching these windows around the other end of the wall).

In other words, the potential reduction of 11 dB.A calculated above for ground floor windows is reduced at window E to 20% of this value, that is, to about 2 dB.A, because of sound reaching the window directly and by diffraction around the end of the screen wall.

Therefore, where walls are used as a defence against traffic noise (a 'line source') they should be returned along the site boundary at least as far as the building line, as shown dotted in the drawing.

In the case of a stationary source of sound (a 'point source'), the wall must be extended on each side of the reception point so that the sound reaching this point by diffraction around the ends is 10 dB.A less than that reaching it over the top. This 'end diffraction' will then not add materially to the sound reaching the reception point over the top of the wall.

Finally, on the question of screening, the reader is reminded that buildings, road embankments, parapet walls, balcony aprons and many other site and building features may serve the purpose of defence against noise.

Aircraft noise

In principle, calculations of insulation requirements for the reduction of aircraft noise are similar to those described above for traffic noise. There are, however, two major difficulties in deciding upon values in dB.A or PNdB as a starting point for calculations

(1) the fact that flight patterns are subject to change, with a considerable effect on noise levels and

(2) the wide range of noise levels since aircraft of various types are heard at very variable distances.

NNI contours do not in themselves provide noise levels. These have to be obtained from the airport authority or by direct measurement over a sufficient period of time for evaluation of the problem. Whether maximum noise level or an average of the higher noise levels is taken as a basis for calculation depends on

*_New Housing and Road Traffic Noise_ (DOE, 1972)

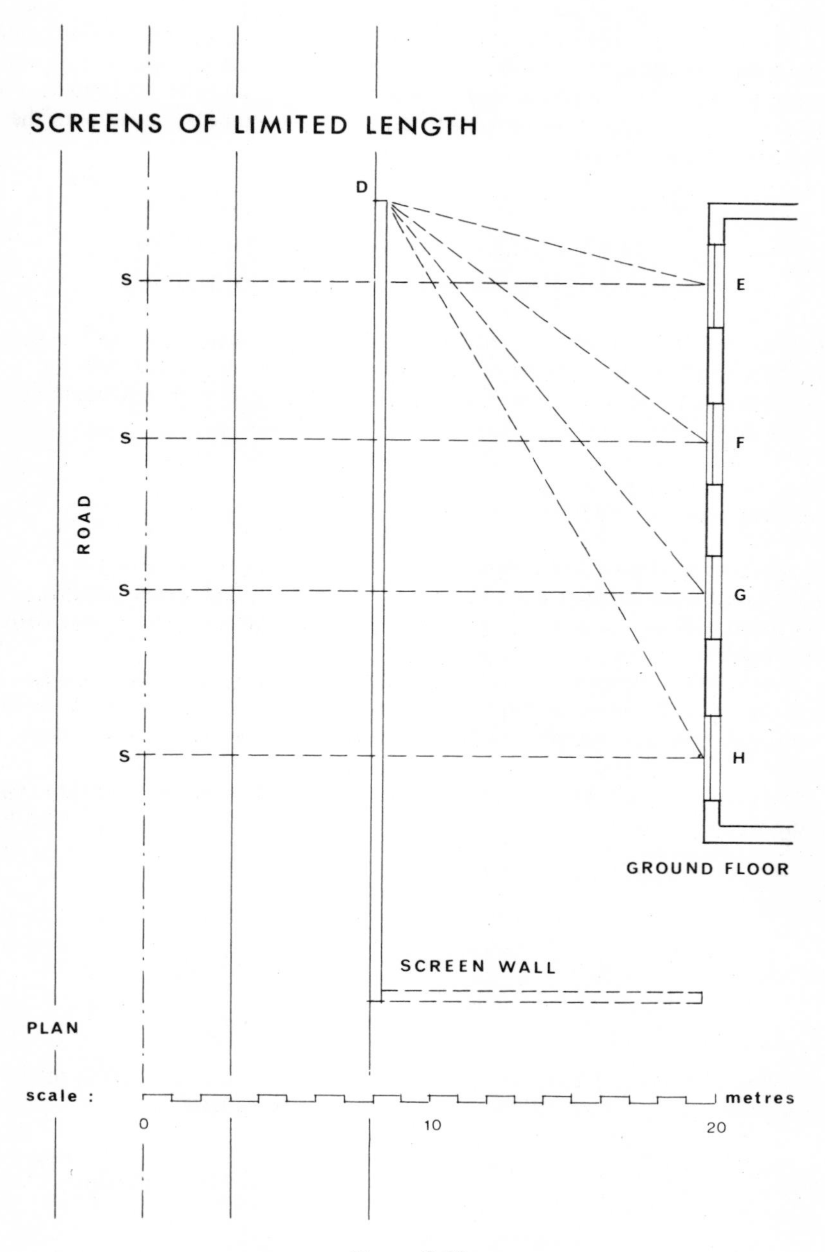
SCREENS OF LIMITED LENGTH
D
S
E
S
F
ROAD
S
G
S
H
GROUND FLOOR
SCREEN WALL
PLAN
scale :
metres
0
10
20

Figure 3.39

the nature of the building being designed. Calculated insulation requirements will of course apply to all façades of the building and the roof.

A fundamentally different approach to the problem is made by the Department of the Environment in their recommendations* to local authorities for approval of housing schemes. This relates insulation requirements directly to NNI values. For example, housing within the contours listed below should be provided with the structural insulation stated

40-44 NNI: 20 dB average insulation (100-3150 Hz)

45-49 NNI: 25 dB average insulation (100-3150 Hz)

50 + NNI: 35 dB average insulation (100-3150 Hz)

It will be seen that this offers a reduction in generalised disturbance rather than providing insulation for specific noise levels and criteria such as are listed on p. 112. In fact it is equivalent to shifting the buildings into a more acceptable contour.

Insulation value of combined structural elements†

It is often necessary to calculate the net insulation value of two structural elements acting in combination. The net insulation of a wall with a window, a partition with a door or a roof with a rooflight, can be obtained by calculation if the insulation values of each element are known.

The method is best explained by an example. Let us assume that the classrooms in figure 3.36 have 6 mm fixed glazing of average insulation 26 dB set in a wall with an insulation value of 50 dB and that the ratio of window to wall is 2:1.

The graph,‡ figure 3.40, may now be used to find the loss of insulation due to the window and thus the insulation value of the two elements in combination

Ratio of areas: 2:1

Difference in insulation: 50 – 26 = 24 dB

From graph, loss of insulation = 23 dB

Therefore net insulation of wall and window = 50 – 23 = 27 dB

If the external noise level (L_{10}) at the classroom windows is 61 dB.A then this will be reduced to 61 – 27 = 34 dB.A intrusive noise level.

**Planning and Noise* (DOE, 1975) which also gives insulation requirements for schools, hospitals, offices, factories and warehouses.

† A six-figure calculation is given in table 3.8 on p. 125.

‡ from *Acoustics, Noise and Buildings*, P. H. Parkin and H. R. Humphreys.

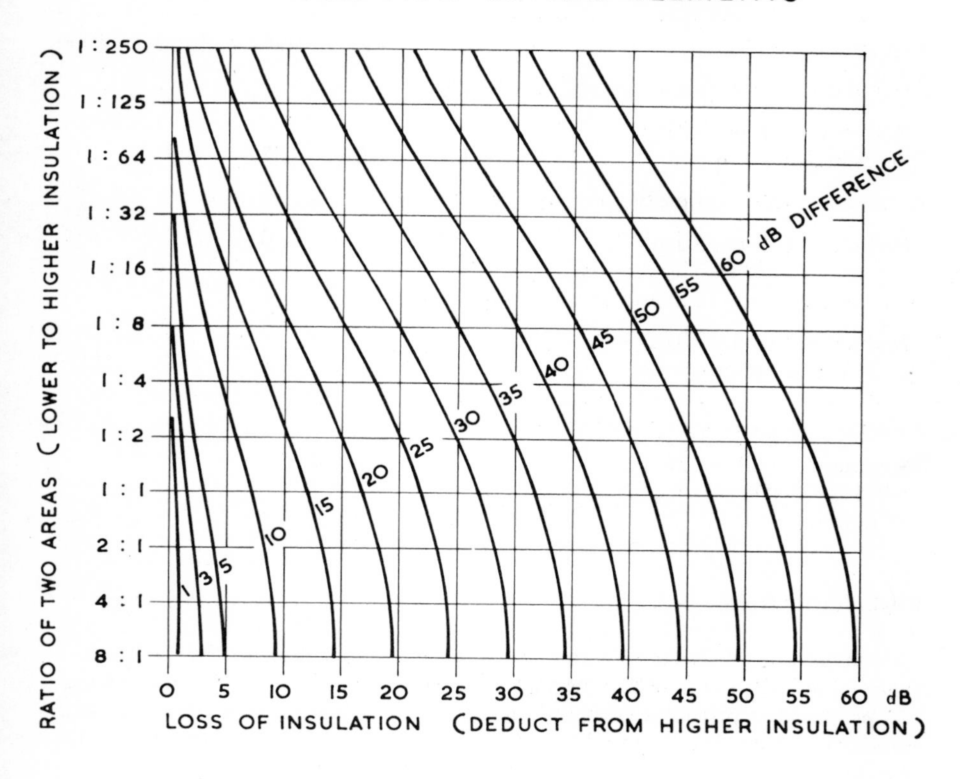

(AFTER PARKIN AND HUMPHREYS)

Figure 3.40

Six-figure Calculations

The preceding calculations are approximate in that the frequency spectra of the offending noises were not related precisely to the insulation values at various frequencies of the building elements used in defence (see tables 3.1 and 3.4). Nor, in the screening calculation, has the change in frequency spectrum which occurs during diffraction been assessed as accurately as is possible. The following calculations therefore employ the available data for

(1) frequency spectrum of the noise (table 3.1)
(2) criterion of acceptable noise (table 3.5) and
(3) insulation values of the structure (table 3.4)

at six octave band centre-frequencies. We also need the following information, previously expressed in dB.A.

TABLE 3.7

Frequency (Hz)	125	250	500	1000	2000	4000
Attenuation due to sound absorption over grass						
Reduction (dB per 30 m) mean height 1.5 m	0.5	1.5	3.0	2.5	1.0	1.0
Attenuation of sound passing through dense foliage						
Reduction (dB per 30 m)	0.8	1.5	1.8	2.0	3.0	5.0
Attenuation of sound due to air absorption						
Reduction (dB per 30 m) at 20°C and 60% humidity	–	–	–	0.1	0.2	0.5
Effect of change in wind direction						
Fluctuation (dB per 30 m) wind speed 10 miles/h plus or minus	0.3	0.5	1.3	2.8	2.3	2.5

Road Traffic Noise

The school previously considered may be taken as an example of the use of six-figure calculations. Let us assume that between the road and the classroom wing it is proposed to provide a lawn and, near the road, a belt of trees 10 m deep with associated shrubs. It is required to determine the type of glazing required for the end classroom to meet a criterion of acceptable intrusive noise of NR 25 (see table 3.5 and figure 3.35).

The calculation is set out in table 3.8 which, in the last column, includes the single-figure calculation for the purpose of comparison.

If 'insulation required' in the table is compared with 'net insulation of window and wall' it will be seen that, at five frequencies, insulation is adequate and is only 1 dB short at 250 Hz. The single-figure calculation does not reveal this, although in this instance the discrepancy is not important. Where however a 'peak' in a noise spectrum happens to coincide with a 'trough' in the insulation curve, the discrepancy could mean a serious weakness in the insulation provided.

In the above calculation air absorption has been included to show how the values compare with the more important reductions. On distances of this order air absorption can safely be ignored.

Effect of screening

Six-figure calculations can also be made to test the effect of a screen between source and reception point. Figure 3.41 provides reduction values in dB in relation to angle of diffraction (θ), effective height of screen (H) and frequency.

TABLE 3.8 Calculation of Insulation Required for Classroom Windows

Mid-frequencies of octave bands (Hz)	125	250	500	1000	2000	4000	dB.A
Road traffic noise (L_{10}) at 10 m from carriageway	75	74	71	65	60	56	**72**
Reduction due to distance of 40 m	8	8	8	8	8	8	**8**
Reduction due to ground absorption, mean height 1.5 m distance 30 m	0.5	1.5	3.0	2.5	1.0	1.0	**1.5**
Reduction due to foliage, 10 m	0.3	0.5	0.6	0.7	1.0	1.7	**1.0**
Reduction due to air-absorption, 30 m	–	–	–	0.1	0.2	0.5	**0.3**
Total reduction	8.8	10.0	11.6	11.3	10.2	11.2	**10.8**
Therefore noise* external to window	66	64	59	54	50	45	**61**
Acceptable intrusive noise (NR 25)	45	37	31	27	23	20	**35**
Therefore insulation required	21	27	28	27	27	25	**26**
Insulation of 6 mm glass in closed windows, ratio 2:1	20	24	28	29	26	30	**26**
Insulation of solid walling	40	44	47	53	57	60	**50**
From figure 3.40, net insulation of window and wall	22	26	30	30	28	32	**27**

*To nearest whole number.

Note that 'effective height' is the amount by which the screen projects beyond the direct sound path, and strictly speaking should be measured at right angles to this path (SR). Reduction due to distance is not included and must be separately calculated for the diffracted path.

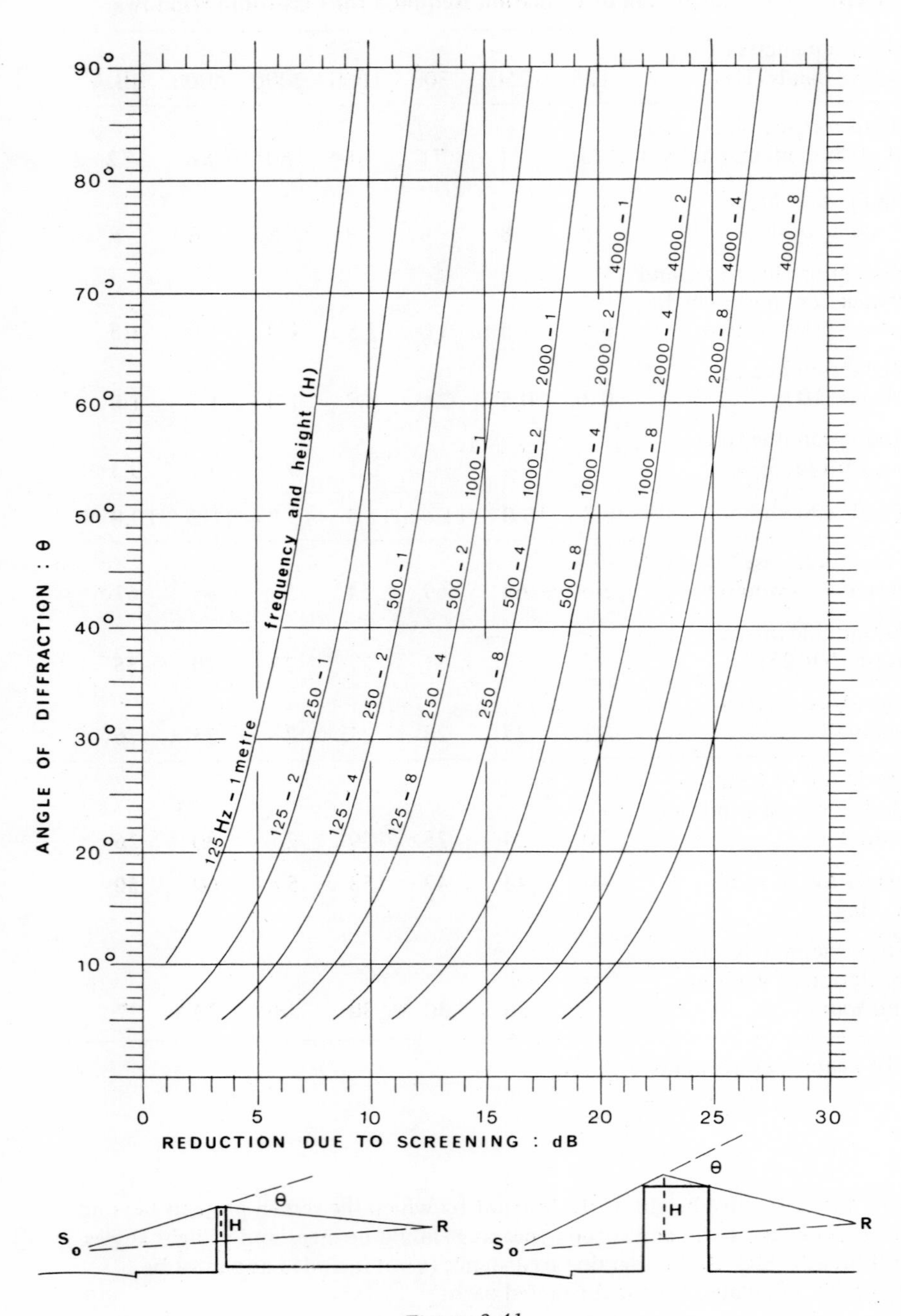

ANGLE OF DIFFRACTION : θ
REDUCTION DUE TO SCREENING : dB
frequency and height (H)
125 Hz – 1 metre
125 – 2
125 – 4
125 – 8
250 – 1
250 – 2
250 – 4
250 – 8
500 – 1
500 – 2
500 – 4
500 – 8
1000 – 1
1000 – 2
1000 – 4
1000 – 8
2000 – 1
2000 – 2
2000 – 4
2000 – 8
4000 – 1
4000 – 2
4000 – 4
4000 – 8
90°
80°
70°
60°
50°
40°
30°
20°
10°
0
5
10
15
20
25
30
S o
H
θ
R

Figure 3.41

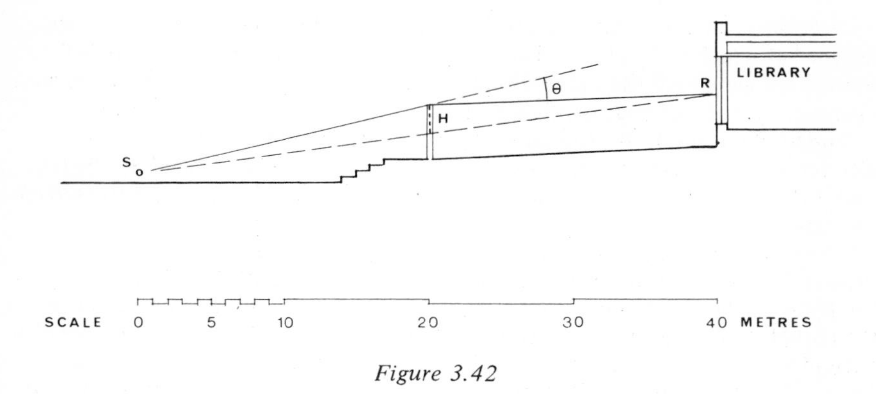

Figure 3.42

The graph in figure 3.41 is based on the formula (due to C. M. Brownsey) $R = 8.2 \log_{10} (44H/\lambda) \times \tan (\theta/2)$, where H is the effective height in metres, λ is the wavelength in metres and θ is the angle of diffraction.

An example will explain the use of the graph. Figure 3.42 shows a section through the land at the rear of a library. At the point S it is proposed to undertake work requiring the use of a pneumatic drill over an extended period of time. It has therefore been decided to erect a temporary screen in the position shown so that intrusive noise in the library does not exceed NR 35 (see figure 3.35).

The calculation is shown in table 3.9. The sequence of the calculation would normally be inverted in order to discover the required height of the screen, but for clarity it is shown in the same sequence as previous calculations.

TABLE 3.9 Reduction in Sound Level Due to a Temporary Screen

Mid-frequencies of octave bands (Hz)	125	250	500	1000	2000	4000
Pneumatic drill noise at 10 m distance (dB)	84	83	79	81	82	88
Reduction due to distance of 40 m	12	12	12	12	12	12
Reduction due to screen $H = 2$ m $\theta = 10°$	3	6	8	11	13	16
Therefore noise level at window	69	65	59	58	57	60
Acceptable intrusive noise, NR 35	52	46	41	36	33	30
Therefore insulation required	17	19	18	22	24	30
Insulation of 6 mm fixed glazing	20	24	28	29	26	30

Note that, unlike road traffic, the source of sound is small in relation to the distance at which its level is measured. It can therefore be treated as a 'point source' reducing 6 dB each time distance is doubled. If, as in this case, the area of operation of the pneumatic drill is limited, the screen need not be unduly extended in its length. It need only be long enough to ensure that the sound diffracted around each end is about 15 dB less than that diffracted over the top. End diffraction will not then add materially to the sound level behind the screen (see figure 3.34).

The screen must of course be imperforate but not necessarily very heavy. Provided the sound transmitted 'through' the screen is 15 dB less than that diffracted over the top, the effectiveness of the screen will not be reduced. Since the reduction due to the screen is 3, 6, 8, 11, 13 and 16 dB at the six frequencies examined, this means that the insulation values of the screen should be at least 18, 21, 23, 26, 28 and 31 dB respectively. This would be provided by a material weighing about 9 kg/m^2.

Reduction of reverberant noise

The value of reducing reverberation in a room by the introduction of special sound-absorbing materials has been discussed under 'Reduction of room noise' on p. 62* and it will have been noted that the extent to which the build-up of reverberation can be reduced is dependent on the *proportional* increase in room absorption, not the absolute increase. Figure 3.43 shows the relationship between

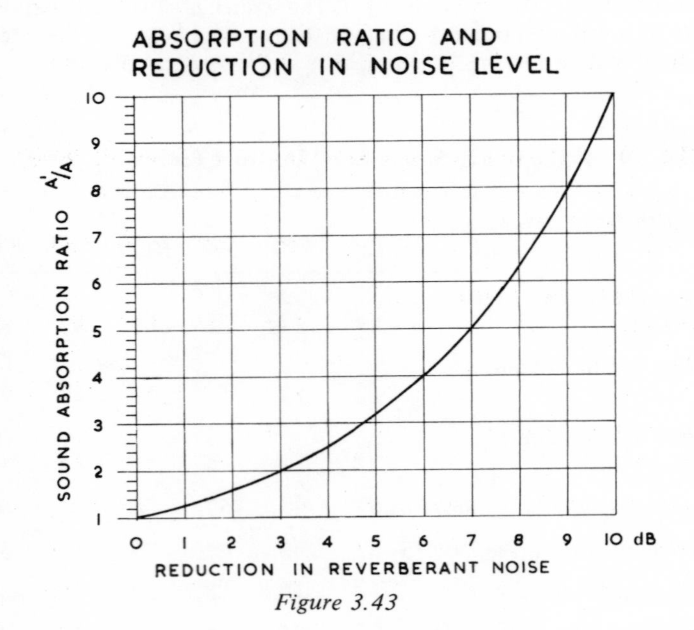

Figure 3.43

*See also figure 4.21

this 'absorption ratio' and reduction in reverberant noise level, the absorption ratio being

$$\frac{\text{room absorption after treatment}}{\text{room absorption before treatment}} \quad \text{or} \quad \frac{A^1}{A}$$

The measurement of sound absorption in square metre sabins has been explained in section 1 (p. 35). The complete calculation is best illustrated by means of an example. Figure 3.44 shows the plan and section of a typing pool in which it is

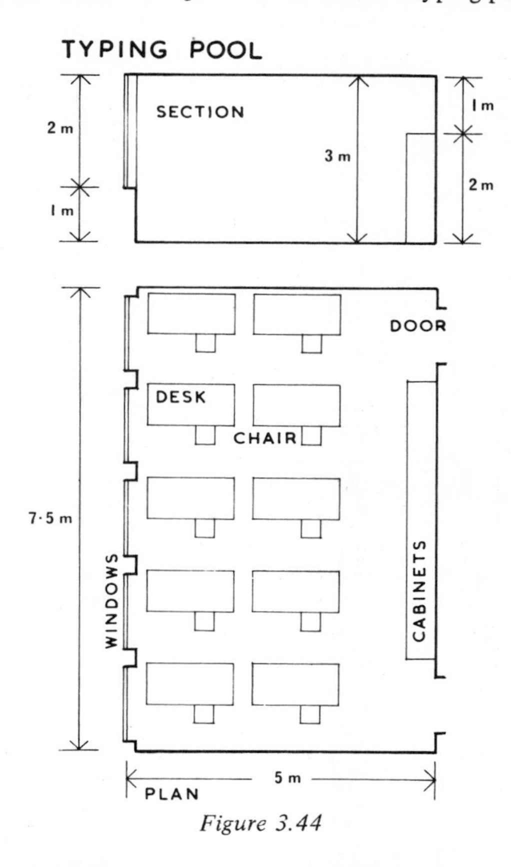

Figure 3.44

proposed to treat the ceiling with special acoustic tiles. Table 3.10 sets out the calculation and it will be seen that at all but the lower frequencies a substantial reduction in reverberant sound level (RSL) is possible. Since the ear is less sensitive to these frequencies, the subjective gain would justify the cost of the special acoustic tiles.

Some account should be taken of the frequency spectrum of the room noise in choosing acoustic materials. Most proprietary materials absorb sound preferentially at middle and high frequencies and deal effectively with frequencies to

TABLE 3.10 Noise Reduction Calculation: Typing Pool

Item and description	Volume, area or number	Coeff. 125 Hz	Metric sabins	Coeff. 250 Hz	Metric sabins	Coeff. 500 Hz	Metric sabins	Coeff. 1000 Hz	Metric sabins	Coeff. 2000 Hz	Metric sabins	Coeff. 4000 Hz	Metric sabins
(1) Air (absorption negligible below 1000 Hz) (m^3)	112.5	–	–	–	–	–	–	0.0033	0.04	0.0066	0.08	0.0165	1.86
(2) Floor, plastic tiles (m^2)	37.5	0.02	0.75	0.04	1.50	0.05	1.88	0.05	1.88	0.10	3.75	0.05	1.88
(3) Ceiling, plaster (m^2)	37.5	0.03	1.13	0.03	1.13	0.02	0.75	0.03	1.13	0.04	1.50	0.05	1.88
(4) Windows, closed, 4 mm glass (m^2)	12.0	0.30	3.60	0.20	2.40	0.10	1.20	0.10	1.20	0.05	0.60	0.05	0.60
(5) Walls, window aprons, piers, plaster (m^2)	51.5	0.03	1.55	0.03	1.55	0.02	1.03	0.03	1.55	0.04	2.06	0.05	2.58
(6) Filing cabinets, metal, surface area (m^2)	15.0	0.20	3.00	0.10	1.50	0.05	0.75	0.03	0.45	0.04	0.60	0.05	0.75
(7) Doors, flush, wood (m^2)	4.0	0.30	1.20	0.20	0.80	0.15	0.60	0.10	0.40	0.10	0.40	0.05	0.20
(8) Desks, wood, number	10	0.20	2.00	0.15	1.50	0.10	1.00	0.10	1.00	0.10	1.00	0.10	1.00
(9) Typists, number	10	0.20	2.00	0.30	3.00	0.50	5.00	0.60	6.00	0.70	7.00	0.60	6.00
Total absorption before treatment, A		15.23		13.38		12.21		13.65		16.99		16.75	
(10 Acoustic ceiling tiles, extra over plaster (m^2)	37.5	0.22	8.25	0.62	23.25	0.63	23.63	0.67	25.13	0.76	28.50	0.70	26.25
Total absorption after treatment, A^1		23.48		36.63		35.84		38.78		45.49		43.00	
Sound absorption ratio A^1/A		1.54		2.74		2.94		2.84		2.67		2.57	
From graph, figure 3.43, reduction in RSL (dB)		1.7		4.2		4.6		4.5		4.3		4.2	

which the ear is most sensitive. Where, however, the noise in question has intense low-frequency components, any proposed treatment must be effective in this part of the spectrum.

The calculation in table 3.10 is relevant to a more or less continuous noise. Impulsive noise of short duration does not build up, yet the time taken for the resulting reverberation to die away is a measure of the acoustic comfort of the room.

We may therefore use the data in the above calculation to determine the 'reverberation time' of the typing pool before and after treatment, reverberation time being the time in seconds for the noise to decay by 60 dB.

Reverberation time (RT) may be obtained by the formula

$$\mathrm{RT} = \frac{V}{A} \times 0.1608$$

where V is the volume of the room in cubic metres and A the total absorption in square metre sabins.

At 500 Hz, for example, reverberation time in the typing pool, without acoustic treatment, will be

$$\frac{112.5}{12.21} \times 0.1608 = 1.5\,\mathrm{s}$$

With the acoustically absorbent ceiling the reverberation time will be

$$\frac{112.5}{35.84} \times 0.1608 = 0.5\,\mathrm{s}$$

At all but the lower frequencies a similar reduction in reverberation time can be expected. This means that impulsive noises, such as the banging of doors or the scraping of a chair, will be cut short and the room will be generally quieter.

Speech interference calculation

The noise rating curves shown in figure 3.35 can also be employed in calculating the insulation required to reduce a noise to a level at which verbal communication is comfortable.

Figure 3.45 gives the suggested noise ratings (NR) appropriate to a range of situations in which it is required to reduce speech interference to an acceptable level. To use this method of calculation it is necessary to know the frequency spectrum of the noise and to decide which of the situations described in the graph is appropriate. The calculation may again be illustrated by an example.

Let us assume that it is required to provide conditions in a foreman's office which will allow conversation to take place at a distance of 3.5 m without raising the voice unduly, and that in the workshop adjoining the office the loudest process has a frequency spectrum as shown in table 3.11. It will be seen from figure 3.45 that intrusive noise should not exceed NR 45. Table 3.11 sets out the calculation and shows that, for example, a wood stud partition with 13 mm dense fibreboard on each side would meet the insulation requirements.

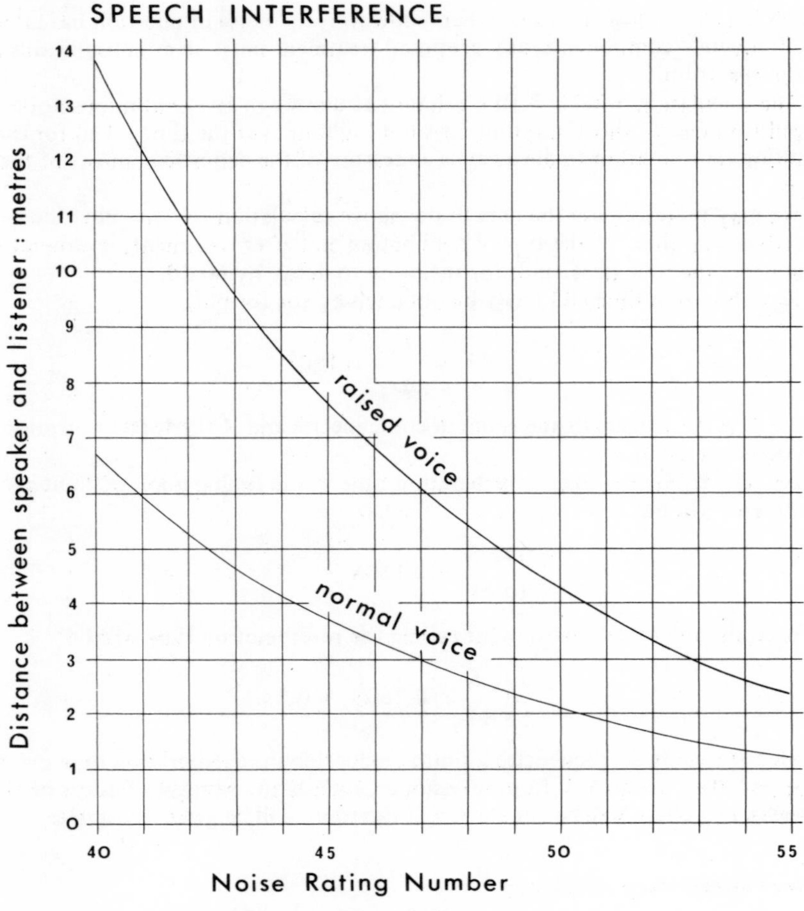

Figure 3.45

If, however, a window is required in the partition the combined elements will have to meet the insulation requirements stated in the table. The calculation for this was given on p. 112. Similarly if, as is likely, a door is required between office and workshop, it should be as heavy per unit area as the partition or, in combination with the partition, give an adequate over-all insulation. All sides of the office, not merely the side facing the noise source, should have the minimum insulation values shown in the table. If the office has an independent roof, as in figure 3.17B, this also should provide the same degree of insulation. Finally, the whole structure must be air-tight, which means that any opening portions should close on resilient beads.

Table 3.5 gave suitable noise rating figures for telephoning. It will be seen that NR 45, employed as a criterion in the above calculation, would be rated as providing 'fair' conditions for the use of the telephone. If a higher criterion is

TABLE 3.11 Speech Interference Calculation

Mid-frequencies of octave bands (Hz)	125	250	500	1000	2000	4000
Workshop noise (RSL) (dB)	72	73	77	84	92	88
Max. acceptable noise levels (NR 45) (dB)	61	54	50	46	43	40
Hence, reduction required by insulation (dB)	11	19	27	38	49	48
Insulation values of wood stud partition with 13 mm dense fibreboard each side (dB)	16	22	28	38	50	52

required there would be two alternatives

(1) the raising of the criterion to NR 40 or
(2) the provision of localised absorbent material around the telephone position.

Note also that if in the workshop there are processes comparable in loudness to, or having markedly different frequency spectra from, the one taken as an example above, then it may be necessary to consider the sources of noise in combination, or separately, to ensure that the insulation of the office enclosure is adequate in all circumstances.

Community reaction to industrial noise

Before siting a factory near a residential area or introducing a new process in an existing factory, some prediction should be made of community reaction to any noise which may be produced. Where a factory introduces a new noise into a neighbourhood complaints and possibly legal action may result.

We have seen that annoyance is much more difficult to assess in advance than, say, interference with speech. Criteria in terms of permissible noise levels are dependent on social surveys and the study of case histories of complaints, related back to the acoustic characteristics of the offending noises.

A number of techniques* have been developed for the prediction of the likelihood of complaints of which the following method based on 'noise ratings' is fairly typical. Essentially, the calculation starts with an octave band analysis of the noise as registered, or predicted, at the nearest dwellings to the factory. This is then given a noise rating number by reference to the graph in figure 3.35. The noise rating curves are considered as 'ceilings' and the lowest of these to 'cover' the frequency spectrum of the offending noise provides the 'basic noise rating'. This rating is then modified by correction factors relative to duration, character,

*See also BS 4142: 1967 Method of Rating Industrial Noise Affecting Mixed Residential and Industrial Areas

time of day, etc. of the noise. From the corrected number, called here the 'annoyance rating', an approximate prediction can be made of the likelihood and extent of complaints. The correction factors and the estimate of community reaction is given in table 3.12.

TABLE 3.12 Industrial Noise: Noise Rating Correction Factors

Type of correction	Description		Correction factors
Duration	Percentage of time	100	0
	that noise prevails	25	– 5
		5	–10
Time of day	Evening		0
	Daytime only		–5
	Night time		+5
Background noise	Very quiet area		+5
	Non-industrial area		0
	Urban area		–5
Nature of noise	Containing marked pure tones		+5
	Impulsive or intermittent		+5
Community attitude	Some previous experience		–5

Estimated community reaction

Annoyance rating	Community reaction
30–40	Few complaints
40–50	Complaints from about half the residents
over 55	Widespread complaints likely to result in legal action

The following is an example of the calculation. Let us take the case of a proposal to site the open end of a saw mill near the boundary of a joinery works in an urban area about 50 m from the nearest dwellings. Let us also assume that the noise from the power saw will reach the levels shown in table 3.13, that it contains marked pure tones and that the saw will be operated intermittently for about 25% of the time during the day in a locality where such a noise will be unfamiliar. The complete calculation is shown in table 3.13.

By reference to table 3.12, it will be seen that the annoyance rating of 65 indicates clearly that the proposed siting of the saw mill would cause widespread annoyance. Full enclosure will be necessary with insulation calculated to reduce the annoyance rating to about 30 if possible.

TABLE 3.13 Community Reaction to Saw Mill Noise

Mid-frequencies of octave bands (Hz)	125	250	500	1000	2000	4000
Noise levels 3 m from saw mill (dB)	71	72	79	79	86	89
Reduction due to distance of 50 m	24	24	24	24	24	24
Hence frequency spectrum near dwellings	47	48	55	55	62	65
Equivalent noise rating number 70	82	76	72	69	67	65

Corrected noise rating		
Basic noise rating near dwellings		70
Duration: saw mill operates 25% of time	−5	
Time of day: daytime only	−5	
Background noise: urban area	−5	
Nature of noise: pure tones	+5	
intermittent	+5	
Community attitude: no previous experience	0	
Total correction		−5
Annoyance rating		65

Sound insulation of party structures

We have seen that, where a noise produced inside a building may disturb people in adjoining rooms, the insulation value required for the dividing wall or floor can be calculated if the level and frequency spectrum of the offending noise is known. In residential buildings of more than one occupancy the party walls or party floors should provide adequate sound insulation but in this case the level and character of the noise to be expected is a variable and largely unknown quantity. Moreover, there will be two types of noise which can be disturbing: air-borne sound through walls and floors, and impact noise through floors.

The method of arriving at a satisfactory acoustic specification for party structures has therefore been to obtain householders' views on the degree of noise disturbance in existing houses and flats of varying construction and then comparing these assessments with the measured insulation values of the party walls or floors. In this way reasonably satisfactory standards of insulation can be established and such standards were embodied in the 1972 Building Regulations.

The measurement of air-borne sound insulation was explained in principle on p. 66. The measurement of the resistance of a floor to the transmission of impact sound is obtained by using a standardised 'impact machine' on the floor in question and then measuring the resulting sound level in the room below. The machine has five hammers operated by a small electric motor to deliver 10 blows per second on the top surface of the floor. Sound intensity levels are taken at various positions in the room below and the results averaged. A frequency analysis is also made of the transmitted noise since, as for air-borne sound, overall sound levels may not provide a sufficiently accurate assessment of the effectiveness of a floor construction. The standardised method of measurement for both air-borne and impact sound transmission is fully described in BS 2750: 1956 Recommendations for Field and Laboratory Measurement of Airborne and Impact Sound Transmission in Buildings.

TABLE 3.14

Frequency (Hz)	Minimum air-borne sound reduction (dB)		Maximum impact sound transmission, party floors (dB)
	Party walls	Party floors	
100	40	36	63
125	**41**	**38**	**64**
160	43	39	65
200	44	41	66
250	**45**	**43**	**66**
315	47	44	66
400	48	46	66
500	**49**	**48**	**66**
630	51	49	65
800	52	51	64
1000	**53**	**53**	**63**
1250	55	54	61
1600	56	56	59
2000	**56**	**56**	**57**
2500	56	56	55
3150	56	56	53

Table 3.14 gives the criteria stipulated by the Building Regulations for the insulation to be provided by party walls and floors between dwellings in separate occupancy or between dwellings and other types of accommodation which may cause noise disturbance. It will be seen that the insulation values and transmission values are given for frequencies at one-third octave intervals between 100 and 3150 Hz. A wall or floor construction is deemed acceptable provided the aggregate of any deviations from these values does not exceed 23 dB.

To avoid undue loss of insulation by flanking transmission the Building Regulations specify in some detail various acceptable arrangements of flanking strucıure, together with appropriate weights per unit area. Subject to these provisions, a number of alternative constructions (including finishes) are described which are 'deemed to satisfy' the regulations. It will be noticed from an examination of these acceptable specifications that, in the case of party walls, weight per unit area, air-tightness and the possible presence of an internal cavity are the main determinants of insulation value. In the case of party floors, weight per unit area and either a soft floor finish or a 'floating floor' are the important determinants of insulation for both air-borne and impact sound.

4 Room Acoustics

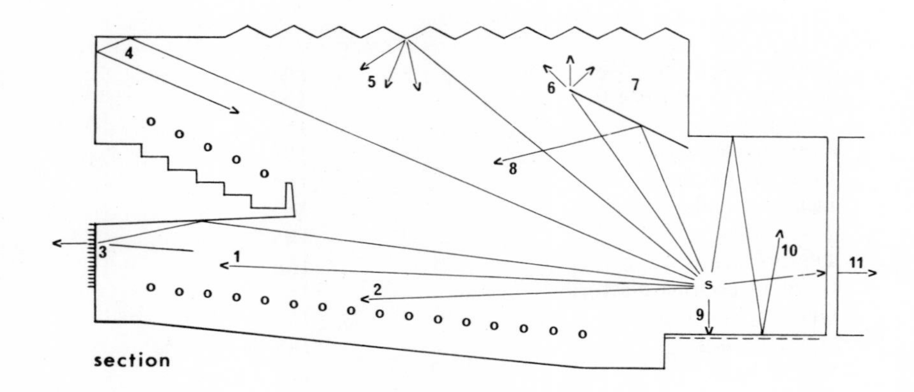

Figure 4.1

Behaviour of Sound in an Enclosed Space

When sound is generated in a room it is transmitted, reflected and absorbed in various ways depending on the shape, dimensions and construction of the enclosure. Figure 4.1 illustrates the various phenomena which may occur and which may be listed as follows with reference to the numbering of the arrows in the drawing

(1) attenuation due to distance
(2) audience absorption of direct sound
(3) surface absorption of direct and reflected sound
(4) reflection from re-entrant angle
(5) dispersion at modelled surface
(6) edge diffraction
(7) sound shadow
(8) primary reflection
(9) panel resonance
(10) inter-reflection, standing waves and reverberation
(11) sound transmission.

The drawing provides a single picture of the phenomena which have been described individually in section 1. All have to be kept in mind in the design of auditoria.

Types of Auditorium

Auditoria can be conveniently classified as follows

(1) for speech
(2) for music
(3) multi-purpose.

Since requirements for the optimum reception of speech differ from those for music, they will be discussed separately and then suggestions will be made for reconciling the sometimes conflicting requirements of speech and music in multi-purpose rooms. Auditoria for speech are, for example, conference halls, lecture theatres and law courts. Auditoria for music range from music practice rooms to concert halls. Town halls and school assembly halls typify the multi-purpose room. Many rooms designed primarily for speech, such as theatres, may on occasion be used for music and then the designer will be required to weigh the relative importance of speech and music in determining a suitable acoustic environment.

Acoustics for Speech

In the design of any auditorium the nature of the source of sound and its location (or locations) have first to be considered. Unamplified speech sounds normally range from about 30 dB.A (whispering) to about 60 dB.A (lecture voice) when measured at a distance of 3 m. An actor may occasionally raise his voice to 70 dB.A but it will be seen that the average voice level, compared with that of an orchestra, is low and that many sounds on which intelligibility depends are very weak indeed.

The other important characteristic of speech is that understanding depends upon the clear reception of a rapid sequence of discrete sounds, some of which are of a very short duration. We may therefore conclude that

POWER + CLARITY = INTELLIGIBILITY

and consider how, in the design of a room for speech, we may optimise conditions for loudness and clarity. The following charts show the factors to be considered under each heading which will then be considered in some detail as they affect design.

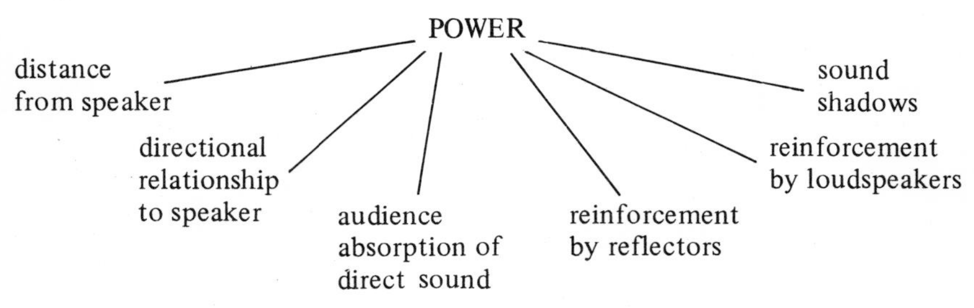

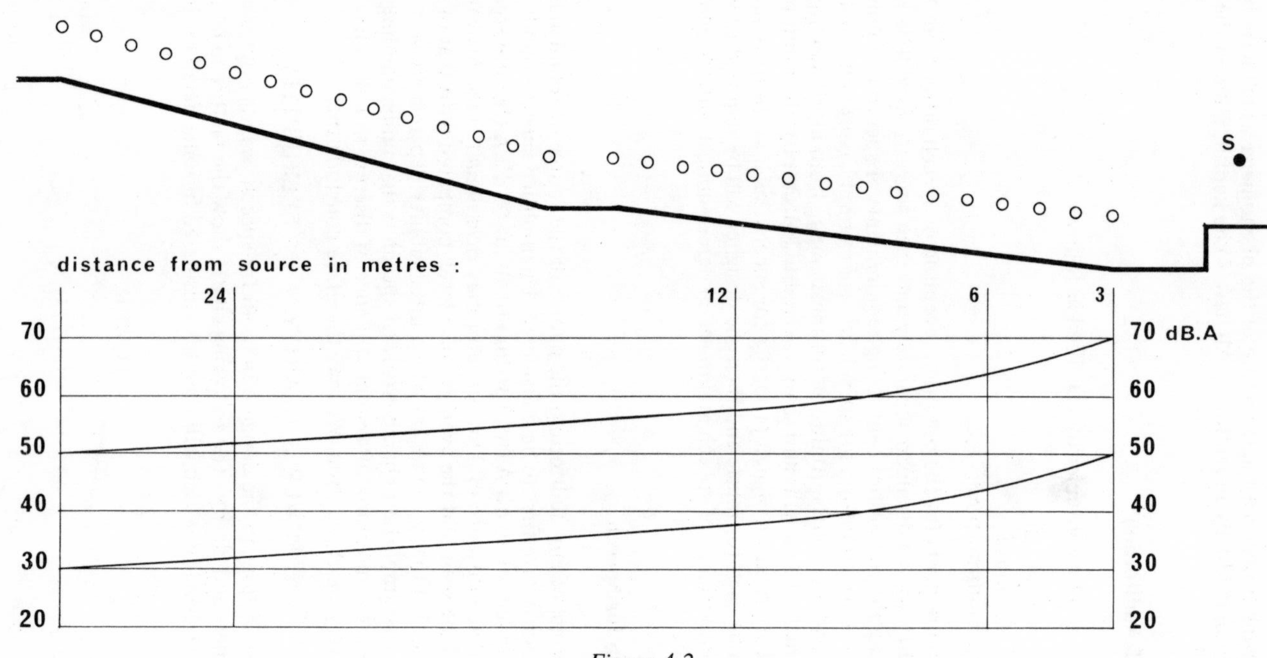
S
distance from source in metres :
24
12
6
3
70
60
50
40
30
20
70 dB.A
60
50
40
30
20

Figure 4.2

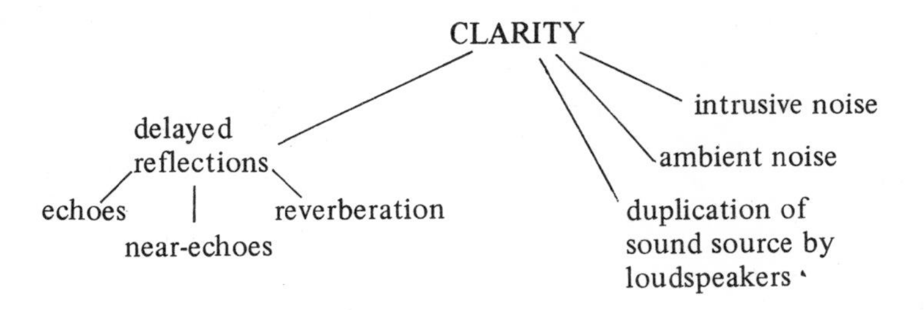

Distance from speaker

Figure 4.2 shows the extent to which the sound of the human voice is attenuated *by distance alone*, in an auditorium with 30 rows of seats. Loud speech (70 dB.A) near the stage falls to 50 dB.A at the rear row. Speech at conversational level (50 dB.A) falls to 30 dB.A, in fact below the level of background noise when an audience is reasonably quiet.

Although the proper placing of sound reflectors may improve this situation to some extent, the importance of keeping the distance to rear rows of seats to a minimum is quite evident. The measures which should be taken are therefore

(1) economy in seat spacing
(2) economy in row spacing
(3) economy in gangway widths within the seating area
(4) economy in number of gangways
(5) optimum shape of audience area (see below)
(6) introduction of a gallery if necessary.

For a given size of audience there are many possible arrangements of seats and gangways, all of which will result in varying distances from the speaker to the rear seats. It is important to discover the arrangement which minimises this distance.

Directional relationship to speaker

Like most sources of sound, the human voice is directional. Knudsen and Harris* found that speech intelligibility varied in accordance with the directional relationship of speaker to listener as shown by the 'equal intelligibility contour' in figure 4.3. The speaker is facing in the direction of the arrow. The contour is given no dimension; at any point on the contour speech intelligibility is equal—equally good or equally poor—depending on its size.

The contour can however be given scale in terms of the following approximate criteria

SA: up to 15 m	=	relaxed listening
15 m to 20 m	=	good intelligibility
20 m to 25 m	=	satisfactory
30 m	=	limit of acceptability without electronic amplification

**Acoustical Design in Architecture*, 1950.

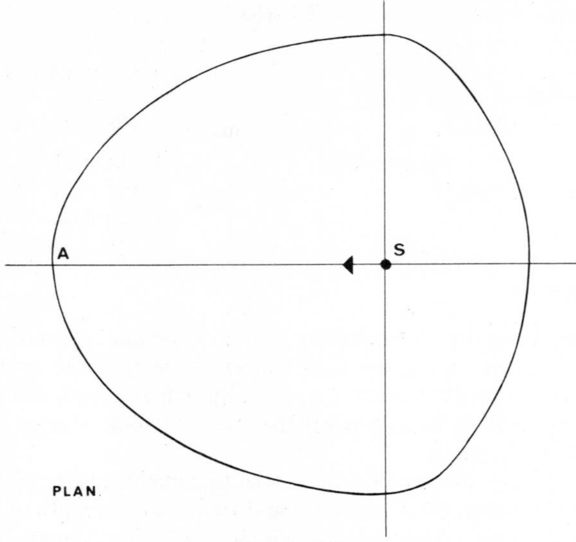

Figure 4.3

Distances on other axes will then be in proportions shown by the contour. The above criteria assume of course that the room is in all other respects well designed acoustically.

If therefore an audience seating area is to be arranged in relation to a fixed position of speaker, who will face his audience for most of the time, then for overall optimum listening conditions it is logical to arrange seating within the contour, or approximately so. From the point of view of hearing alone, positions behind the speaker but within the contour will be better than positions in front of the speaker but beyond the contour. Only because of sight lines will seats behind or to the side be excluded.

There are, however, many situations in which the speaker, or speakers, do not take up one position in relation to the audience. The contour can still be used, however, to find a sensible arrangement of seats. One example, that of the theatre, may here suffice to show how the contours can be used.

Figure 4.4A shows the plan of the stage of an 'open stage theatre'. During performances in this type of theatre actors tend to face each other more often than is the case of the 'proscenium arch theatre'. This results in difficulty in hearing at times when an actor has his back turned to a section of the audience at the side of the stage, if they are too remote from the speaker. The drawing shows positions actors take up *for most of the time* in this kind of production. By rotating the contour from three typical positions, as shown, we may conclude that the optimum arrangement of seats will lie within the shaded area embraced by all three contours, and within appropriate sight lines to the stage.

Three other examples of the application of the contour are given in figure 4.4, in relation to the 'box set' stage, the Restoration apron stage and theatre-in-the-round. For a given *acoustic* criterion the differences in audience capacity become evident although *visually* the box set type of production is superior.

It will be found that the 'equal intelligibility contour' can assist in the design of most auditoria for speech and it will be seen that the traditional oblong shape for halls has little justification as far as speech acoustics is concerned.

Finally, the reader may care to observe how a crowd arranges itself around an open-air speaker when sight lines are of little importance compared with the need to hear. Given the space, the crowd assumes roughly the shape of the contour in figure 4.3.

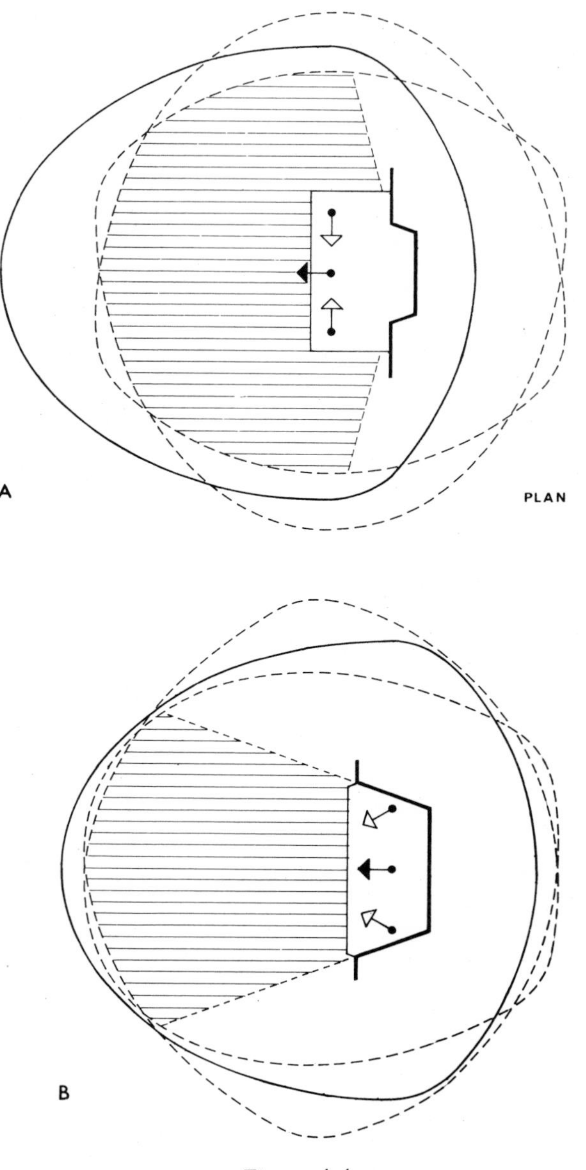

Figure 4.4

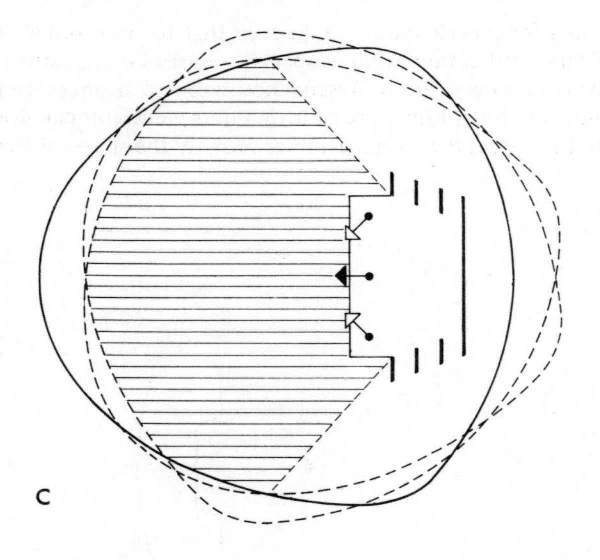

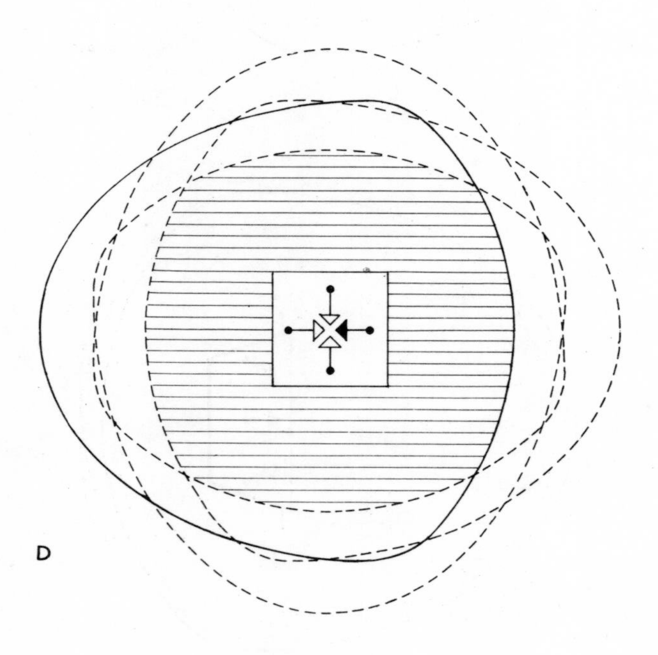

Figure 4.4 (contd.)

Audience absorption of direct sound

Direct sound waves passing at grazing incidence over an audience are absorbed rather in the manner indicated by the arrows in figure 4.5A. This absorption is cumulative and is chiefly responsible for the difficulty experienced in hearing at the rear of a hall with a level floor. This phenomenon causes a reduction in sound level *in addition to* the reduction due to distance.

If, however, we consider the case of an audience seated on a raked (stepped) floor as in figure 4.5B, we will see that the part of the initial sound wave reaching listeners in the rear row is well above the heads of listeners in the front rows and therefore subject to little or no absorption. P and P′ indicate the parts of the sound waves referred to. Thus audience absorption of direct sound can be reduced, and even eliminated, by raking the floor of an auditorium.

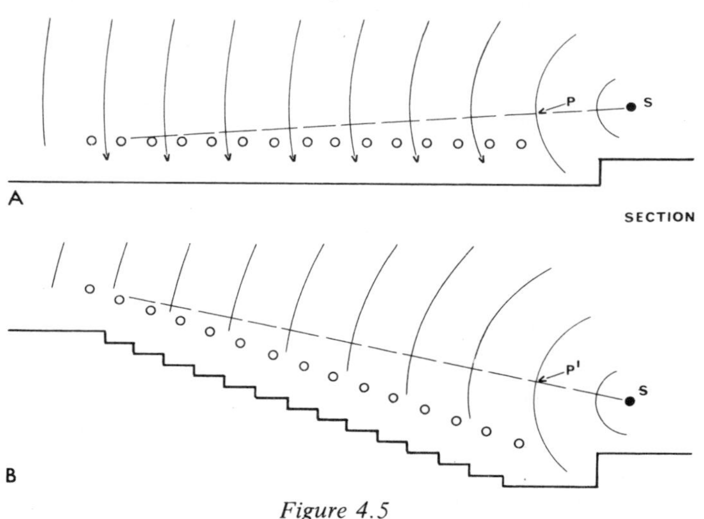

Figure 4.5

Experiments measuring the reduction of 'white noise' passing over the heads of an audience suggest that the rake should be of the order of 20° to eliminate 'audience absorption' in an auditorium with about thirty rows of seats.*

The results of these experiments are shown in the graph (figure 4.6) and clearly indicate that in extreme cases of a level floor and a low platform the reduction in sound level at rear rows due to 'audience absorption' may be as much as that due to distance.

It should be noted that, although galleries are raked for sight lines, sound waves may still pass at grazing incidence over the heads of the audience owing to the relationship between sound source and the angle of the gallery. It follows therefore that the rake of the gallery should be as great as possible, consistent with safety, and that its depth should be limited.

*'White noise' is sound with a balanced mixture of frequencies over a wide range and can therefore be roughly equated with the range of speech frequencies.

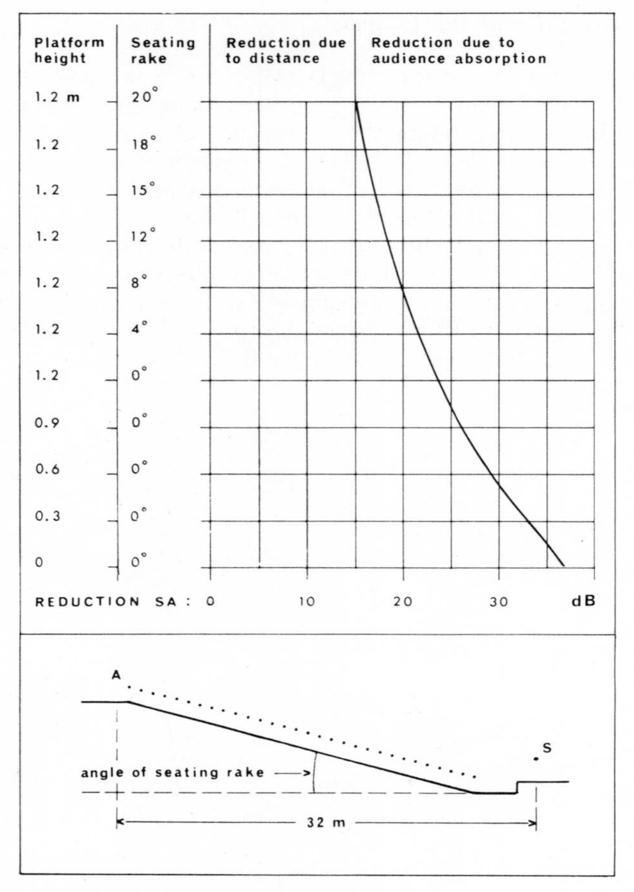

Figure 4.6

Reinforcement by reflectors

Reflectors, if correctly placed, provide a small but significant reinforcement of direct sound. In effect, the listener hears two sounds superimposed (direct and reflected) which, if the time interval between them is short, will result in a raising of the sound level subjectively. In the case of sounds of multiple frequency, such as speech, it is unlikely that this will result in an increase of more than 2 or 3 dB. Nevertheless this may make all the difference between hearing and not hearing very weak speech sounds.

If, by force of circumstances, the floor of an auditorium has to be level, or nearly so, the sound level by way of the reflector may well be higher than that received by direct path because the former will not have been attenuated by 'audience absorption'. Overhead reflectors therefore become more important in this situation than if the floor is well raked, but are an advantage in all situations provided the following design principles are observed

(1) the reflector should preferably be overhead (part of the ceiling or suspended) so that reflected sound is not reduced by 'audience absorption' as is often the case with wall reflectors

(2) the reflector should be as low as possible to reduce the time interval between reception of direct and reflected sound to a minimum

(3) it should be positioned, or angled, so that the rear rows of the audience benefit

(4) it should be wide enough to reflect sound across the full width of the rear row of seats

(5) its minimum dimension should be of the order of 3 m so that the reflected sound is not materially weakened by edge diffraction, as is the case when the edges are too close to each other

(6) it should be flat, or very nearly so (a convex curve attenuates the reflected wave)

(7) it should have a low coefficient of absorption at all frequencies.

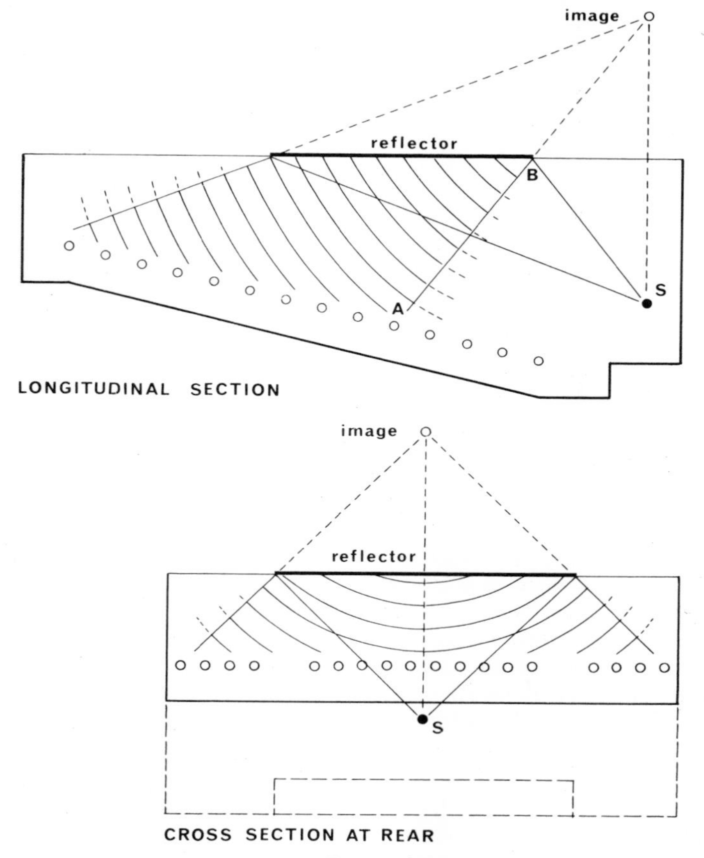

Figure 4.7

Factors (2) and (3) can be studied on the drawing board using the simple geometry of reflection described in section 1. Figure 4.7 shows the simplest case of part of the ceiling of an auditorium employed as an overhead reflector. Note that the reflector is wide enough to project sound over the rear row of seats, shown in cross section, and that it is assumed that the source of sound is central on the platform. The diffraction of the sound waves is shown by dotted lines. The extent of diffraction will vary with frequency but for the more important speech sounds of high frequency this will be small and will not materially reduce the power of the main waves.

It remains to check the difference in sound path between direct and reflected sound, that is, SBA – SA. This should not be more than 11 m–equivalent to a time delay of 1/30 s–and preferably not more than 7 m. Figure 4.8 shows the relationship between such sound path differences, the equivalent time delays, and their effect on speech clarity.

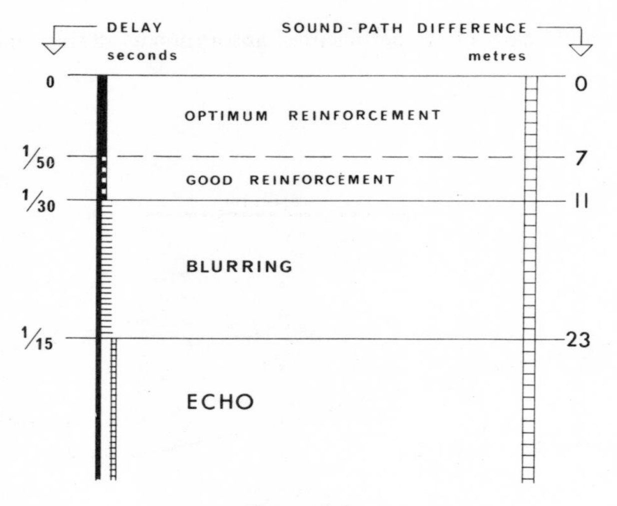

Figure 4.8

The source of sound is, however, seldom in one fixed position and figure 4.9 shows the geometry required to provide a sufficiently large reflector for a theatre where the movement of speakers is considerable. The required size of reflector is worked out for each extreme position: S1, S2 in the longitudinal section and S3, S4 in the cross section (figure 4.10), resulting in a reflector ADGF to meet the over-all requirements of sound sources within the acting area.

Where the sound source may be anywhere in the room, as in a council chamber, the reflector should be above all positions of source as shown in figure 4.11. In this drawing extreme positions for speakers (S1 and S2) are examined and it will be seen that the reflecting surface ACFD is large enough to provide reinforcement of sound at all listening positions in respect of sources between the extremes. The remaining parts of the ceiling can be treated with sound-absorbing material if required for the reduction of reverberation.

A reflector need not of course be part of the ceiling. In large auditoria requiring fairly high ceilings this may result in an excessive delay between reflected and

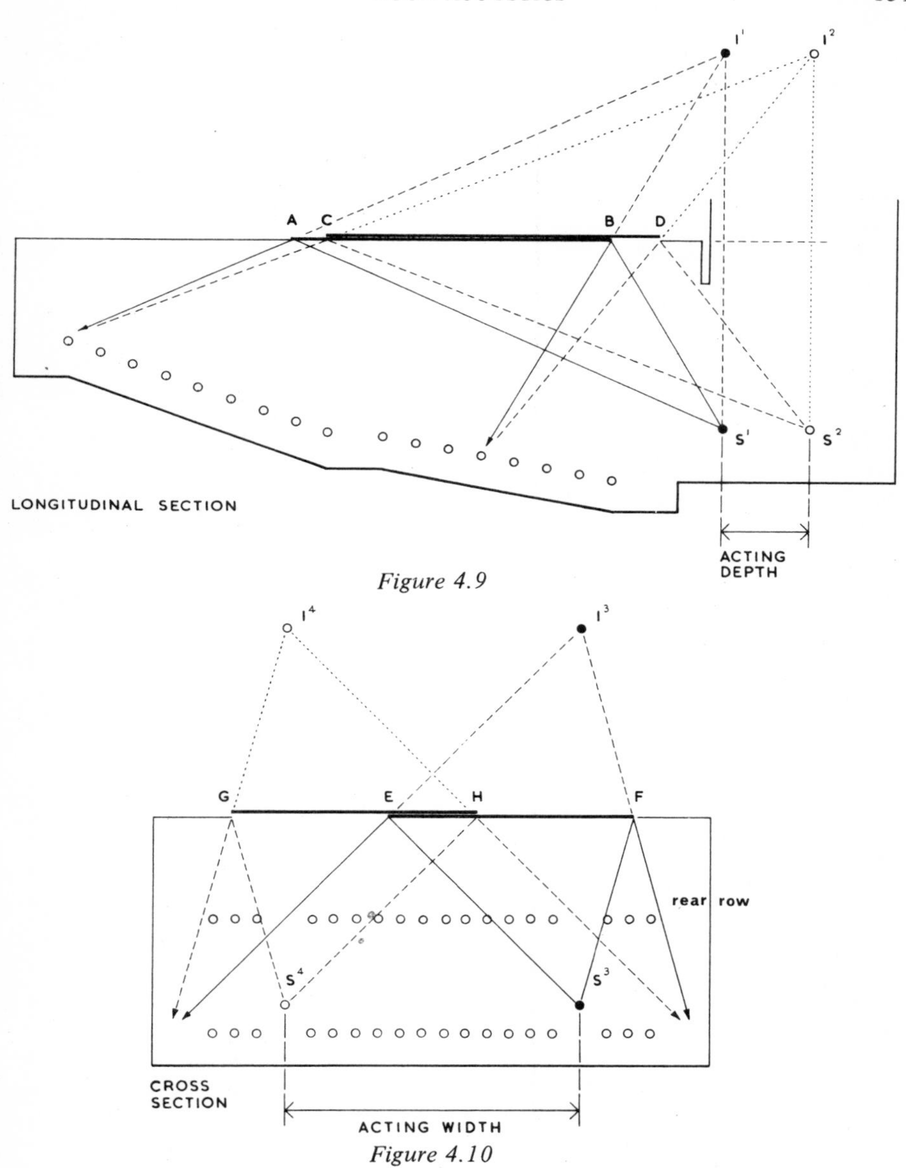

Figure 4.9

Figure 4.10

direct sound. Reflectors suspended (either horizontally or at an angle) are frequently employed at a lower level just in advance of the source of sound. Such a reflector is shown in figure 4.12. This drawing also shows two additional reflectors which, because they are further from the source, need not cause an excessive delay because they operate for seats having a longer direct path. The main value of these additional reflectors is, however, to provide extra reinforcement, progressively, for seats further from the stage. Sound rays, rather than sound waves, are here shown so that the overlapping of the reflections is clear.

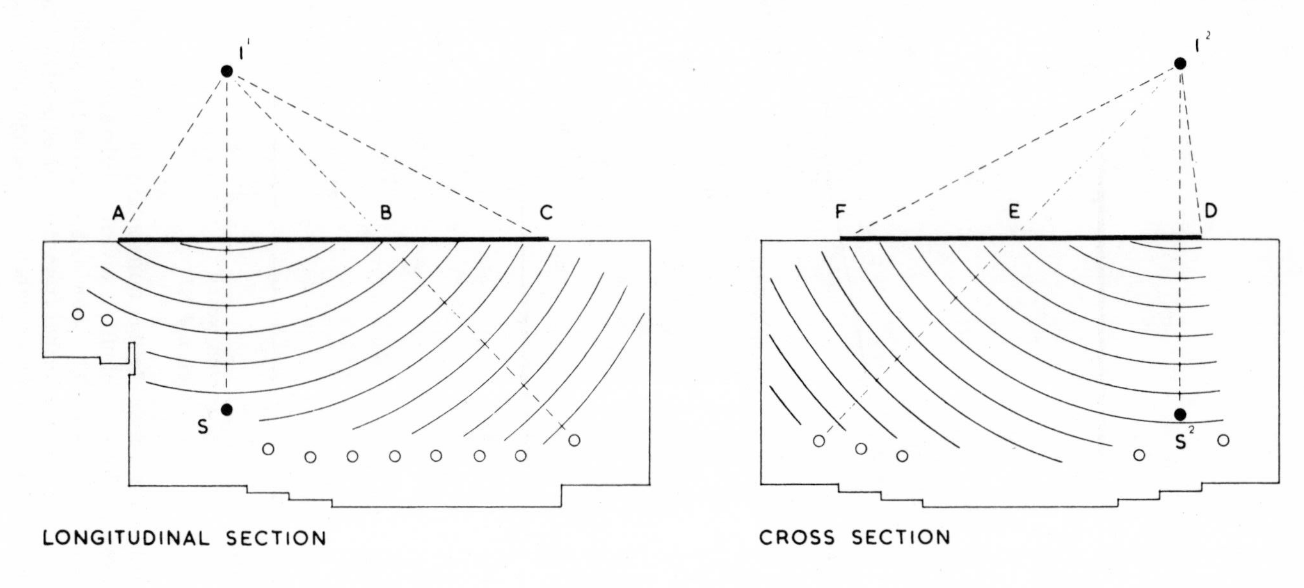

Figure 4.11

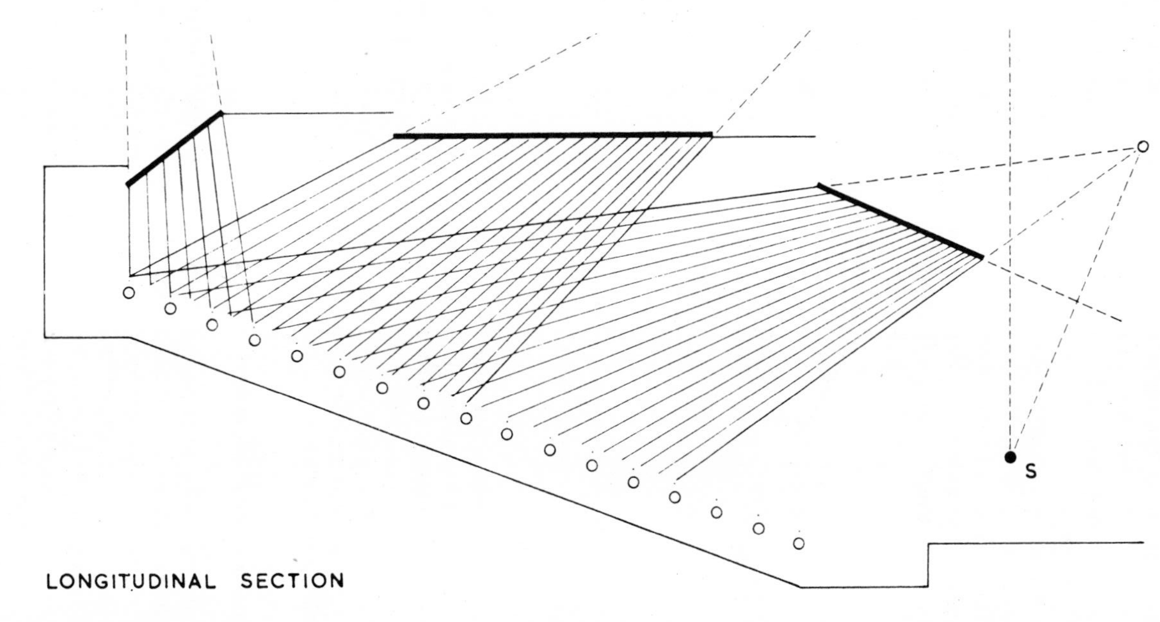
S
LONGITUDINAL SECTION

Figure 4.12

So far, all the reflectors shown have been flat. If for aesthetic reasons it is desired to make them curved it should be remembered that reflections from a convex surface are weaker than from a flat or concave surface (see p. 24). Therefore the curvature should be very small. On the other hand, a concave surface will strengthen reflections, but the effect of such curvature must be carefully examined on the drawing board (using the geometry explained on p. 26) to ensure that this does not result in localising the reflections to the neglect of other areas of seating.

This effect is shown at its worst in figure 4.13 where the ceiling is excessively curved in cross section so that only the centre seats receive reinforcement when the speaker is in a central position on the platform. What is perhaps worse is that, as the speaker moves laterally, this reinforcement will swing from side to side across the auditorium, causing a fluctuation of sound level.

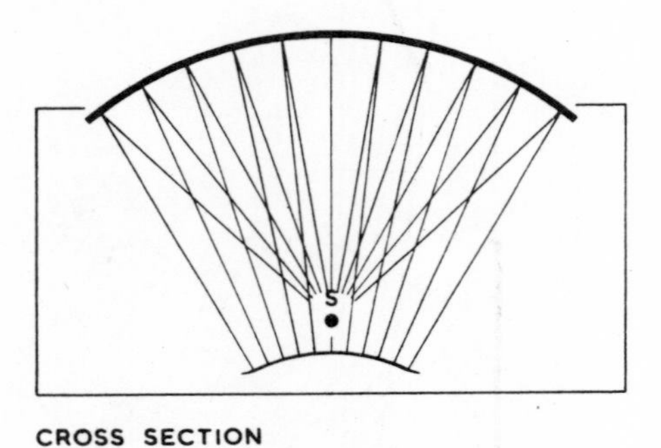

Figure 4.13

Although, generally, overhead reflectors are more effective than vertical reflectors, there is one situation where a vertical reflector is useful. This is in the lecture theatre where frequently a lecturer may stand to one side of the dais and speak while lantern slides are being projected. He is then likely to have his back turned to part of his audience. The flanking reflecting surfaces shown on plan in figure 4.14 can then usefully reflect sound to those behind him, provided the time delay is not excessive.

Reinforcement by loudspeakers

Speech can of course be amplified electronically by means of microphones and loudspeakers and, in large auditoria, this may be the only way in which to meet the requirement of adequate loudness. It must, however, be stressed that the provision of electronic amplification does not obviate the need for good acoustic design of the interior. In fact, faults in design may be exacerbated owing to the increased power available and the duplication of the sound source.

On p. 143 approximate distances at which speech is intelligible were given *on the axis of the direction the speaker is facing*. This may be used as a guide to the need, or otherwise, for introducing loudspeakers, provided the 'equal intelligibility contour' is used for distances other than on the central axis. Some reduction on

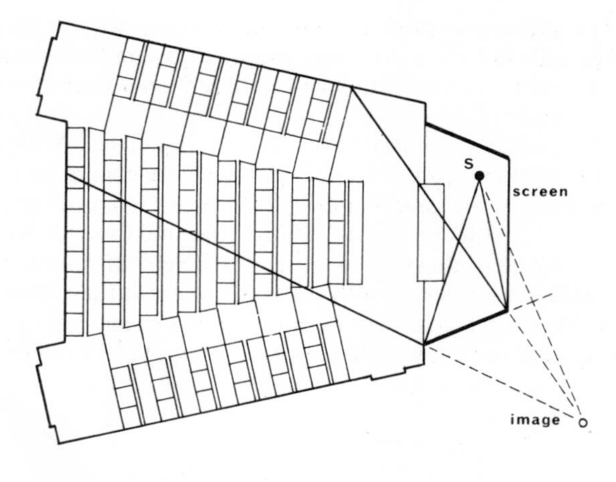

Figure 4.14

these distances may have to be made in cases where inexperienced speakers are expected to take part in the proceedings, such as at a conference. The large conference hall is in fact an example of the need for electronic amplification, not necessarily because of the size of the room but because speakers in the body of the hall may speak with their backs turned to a large proportion of the delegates. The provision of microphones, either in gangways or near the platform, reduces the tendency to speak from wherever the delegate may be sitting.

Ideally, loudspeakers should be used to raise the level of speech in only those parts of the auditorium in which, by distance and possibly audience absorption, the sound has fallen to an unacceptable level. Since loudspeakers are directional, especially at the more important high frequencies, this is possible. The added level of sound should only be sufficient for good intelligibility, otherwise realism suffers.

In very large or lengthy auditoria, such as cathedrals, loudspeakers may be distributed down the length of the interior serving sections of the congregation and operating at a lower level than would be possible with only one loudspeaker (or one cluster) near the pulpit.

Most of the problems arising from the use of loudspeakers are a matter of loss of 'clarity' and therefore will be discussed under that heading (see 'Duplication of sound source by loudspeakers', p. 165). There are, however, certain practical considerations such as initial cost, maintenance, expert operation, restriction of movement for the speaker and the convenient positioning of microphones which suggest that natural acoustics should be exploited to the maximum by good design before the decision is made to introduce loudspeakers.

Sound shadows

Under the general heading of 'power' there remains one final possibility: that it may be reduced because of an obstruction in the path of sound between speaker and listener. The author can think of only one situation where this may occur (in a conference hall) but there are many examples where theatres have deep balconies which prevent the reception of useful *reflected* sound by those seated below.

If we take the case of the conference hall first, figure 4.15A shows the situation where a delegate speaking from his seat under a gallery cannot be heard, by direct sound path, by those in the gallery, nor by reflections from the ceiling. Those in the gallery will be dependent on weak diffracted sound over the balcony apron. At the back of the gallery this will be quite insufficient. It follows therefore that, in those cases where delegates are allowed to speak from their seated positions (almost impossible for the chairman to prevent), balconies should not overhang the main seating area and a low reflecting ceiling should be provided as in figure 4.15B.

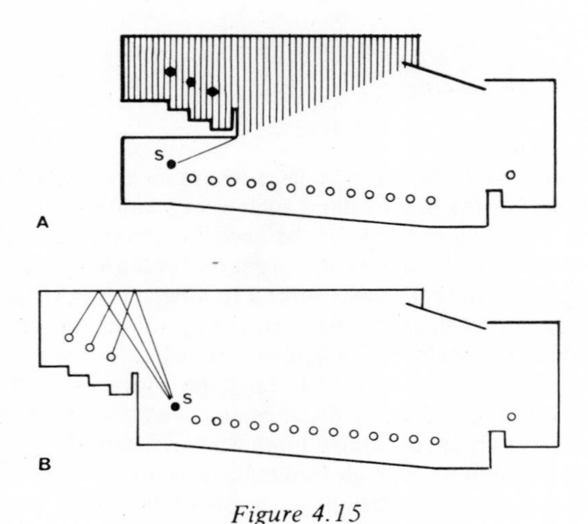

Figure 4.15

Figure 4.16A shows a section through a theatre auditorium in which people below the gallery are prevented from receiving useful reinforcement of sound by ceiling reflections because they sit 'in the shadow' cast by the balcony. Of course the shadow is not as precisely defined as the diagram suggests but nevertheless, at the rear, diffracted sound will be very weak. The section shown in figure 4.16B, having the same number of rows of seats, would be better even though some rows are further removed from the stage. If, however, a low,angled reflector is used (shown dotted), reflected sound may be projected under the gallery.

Delayed reflections: echoes

We now have to consider the second major requirement for speech intelligibility: clarity. On pp. 148 to 150 the value of reflected sound in the reinforcement of speech was described but it was also indicated that, if the time interval between

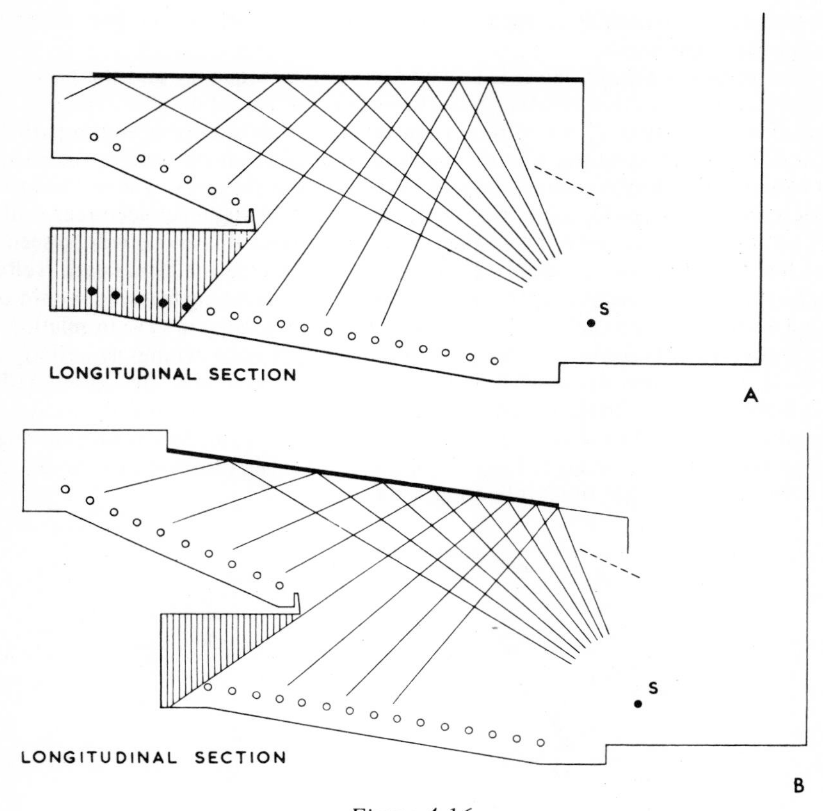

Figure 4.16

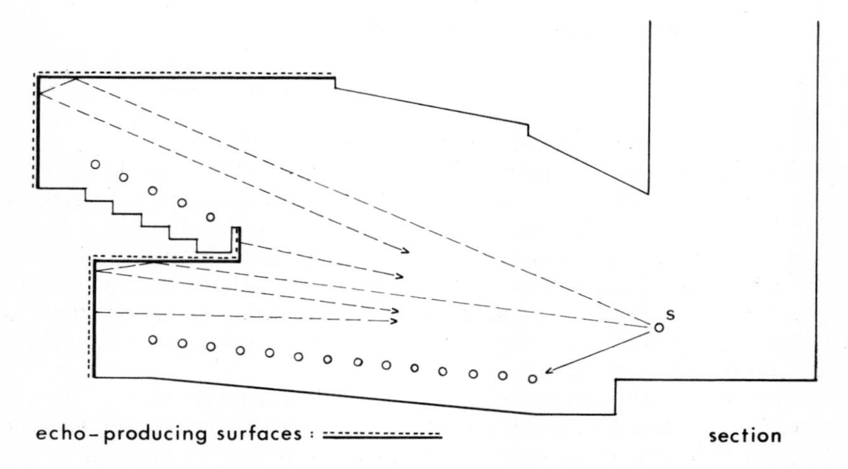

Figure 4.17

the reception of sound by direct path and by reflected path is excessive, disturbing echoes might occur.

Perceptions of echoes depends to some extent on frequency and the power of the reflection in relation to other reflections, but the risk of audible echoes is present if the delay is 1/15 s or more (equivalent to a difference in sound path of 22 m or more). In auditoria where the source of sound is at one end of the room, for example the theatre, echoes may be heard in the front rows of seats because of delayed reflections from the rear wall, the re-entrant angle between rear wall and ceiling, or the balcony apron. Figure 4.17 illustrates this and it will be seen that the delay time can be simply calculated by comparing the direct path (solid line) with the total indirect path to and from the reflecting surfaces. It should be noted also that, if any 'echo-producing' surfaces are curved, concave in relation to the source, echoes will be more powerful. This applies to ceilings as well as walls, as illustrated in figure 4.18. In all cases such surfaces must be treated with sound-absorbing material, or a combination of absorbent material and a modelling of the surface. In the case of the potential echo from the 90° re-entrant angle such treatment must be carried right into the corner. High-frequency echoes can be detected from quite small reflecting surfaces.

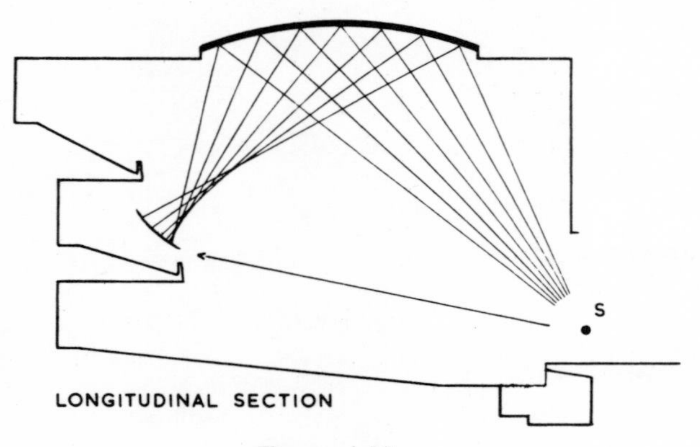

Figure 4.18

The source of sound is, however, not always at one end of the room. In a conference hall or council chamber it may be anywhere in the room. In this case any wall surface or re-entrant angle may produce echoes. Sound path differences must be measured for a number of possible positions of source and it may be necessary to treat all vertical surfaces with absorbent material. An example is shown in figure 4.15, which shows a section through a conference hall in which the wall behind the platform is a potentially echo-producing surface. Similarly, in a wide conference hall, side walls can produce echoes in relation to a source of sound near the opposite wall.

The treatment for the reduction of echoes will depend on the *form* of the surface. As explained on p. 24 the reflection from a concave surface will be more powerful than from a flat surface but from a convex surface less powerful. Modelling of the surface will, however, in all cases reduce, subjectively, the strength of the echo, by dispersion. As an approximate guide to the treatment of echo-producing surfaces the following figures provide safe standards

Flat surfaces: 70% absorbent or 50% with bold modelling
Concave surfaces: 90% absorbent or 70% with bold modelling
Convex surfaces: 50% absorbent or 30% with bold modelling

Delayed reflections: near-echoes

Between a reflection which effectively reinforces speech sounds and one which causes a distinct echo lie reflections which, subjectively, *extend* the direct sound and cause 'blurring'. These are reflections arriving about 1/30 to 1/15 s after the direct sound and occur therefore when sound path differences are between 11 and 22 m.

Again, the effect is frequency dependent and related to the incidence and power of other short-delay reflections. It is, however, advisable to reduce these 'medium-delay' reflections (which we may call near-echoes) by the modelling of surfaces which can produce them. Typically, such delays are caused by surfaces flanking or well above the source of sound, as shown in figure 4.19. In any case, such surfaces contribute nothing to the reinforcement of sound and, as will be seen later, are better employed for the diffusion of reverberation. Alternatively these surfaces can be angled to project sound to where it is most needed at the rear of the room.

Delayed reflections: reverberation

The phenomenon of reverberation has been explained in section 1. We are here concerned with its effect on the clarity of speech. Figure 4.20 shows in the form of a simplified graph the discrete speech sounds of two words plotted against loudness and time.

The duration and power of each sound, as articulated, is shown by the heavy horizontal lines. The extension of each sound by reverberation is shown by the sloping lines. It is evident that the longer the time of reverberation, the greater will be the masking of one sound by the reverberation of the preceding sound, or sounds. This will in particular be the case where a weak sound (most consonants) is preceded by a more powerful vowel. Figure 4.20 could be re-drawn for an open-air situation (non-reverberant) by omitting the sloping lines altogether, when no masking would occur. Alternatively, in a more reverberant situation, such as a cathedral, the sloping lines representing the decay of rever-beration would be at a bigger angle and masking would be extended. Notice also that reverberation tends to fill in the very short time intervals between words upon which speech intelligibility partly depends.

The general effect of reverberation, if excessive, is therefore to blur speech so that the control of reverberation is perhaps the most important factor in the acoustic design of a room.

The time taken for the reverberation of a sound to decay to inaudibility will depend on

(1) the initial power of the sound

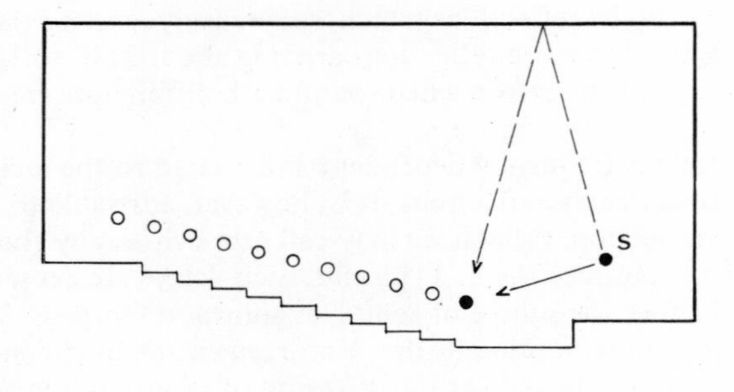

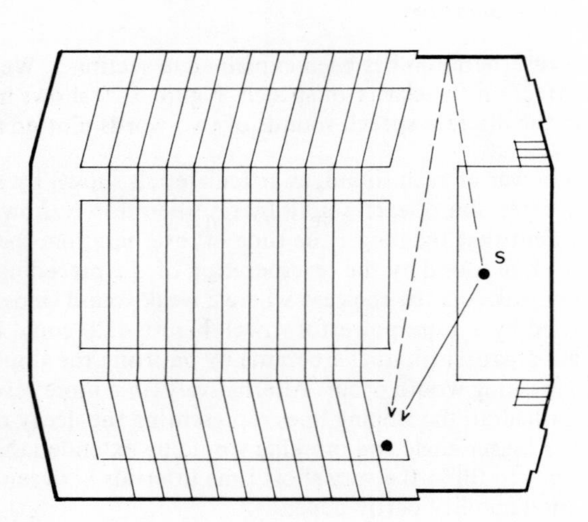

Figure 4.19

(2) the absorbency of surfaces or objects with which it comes into contact during inter-reflection
(3) the volume of the interior and therefore the length of the sound paths
(4) the presence of any standing wave phenomena and
(5) the varying sensitivity of the ear at different frequencies.

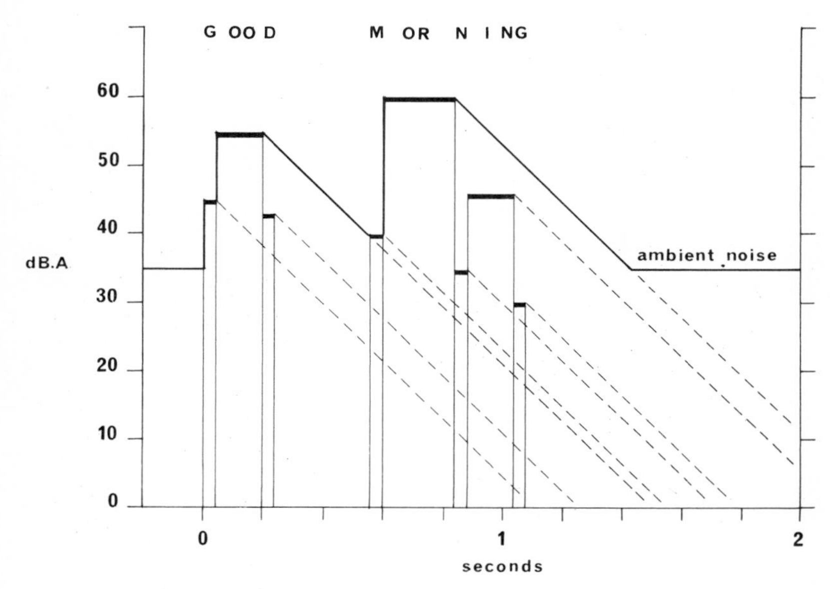

Figure 4.20

Subjectively, the greater the power, the greater the volume and the higher the frequency, the longer will reverberation be apparent. The greater the total absorption of the room, the shorter will be the reverberation. Standing waves will have the effect of extending reverberation for certain frequencies only.

In order to establish criteria for acceptable degrees of reverberation for various purposes (for example, speech and music) the *rate of decay* of reverberation is stated rather than any absolute time of reverberation, since the latter will vary in any given situation as the power of the initial sound fluctuates.

Thus by general agreement 'reverberation time' is defined as the time taken for a sound to decay by 60 dB. This will be more clearly seen in figure 4.21, which shows graphically the decay of a sound of 85 dB to a level below audibility. The reverberation decay 'curve' shows that the rate of decay is 60 dB in 2 s. By definition therefore 'reverberation time' is 2 s, regardless of the power of the initial sound. This enables us to compare the acoustic environment of one room with another and establish criteria. In fact the environment represented in figure 4.21 is only moderately reverberant. A large church might have a reverberation time of 5 s or more but, for the clear reception of speech, the decay curve should be steeper, with a reverberation time of 1 s, as shown by the broken line on the graph.

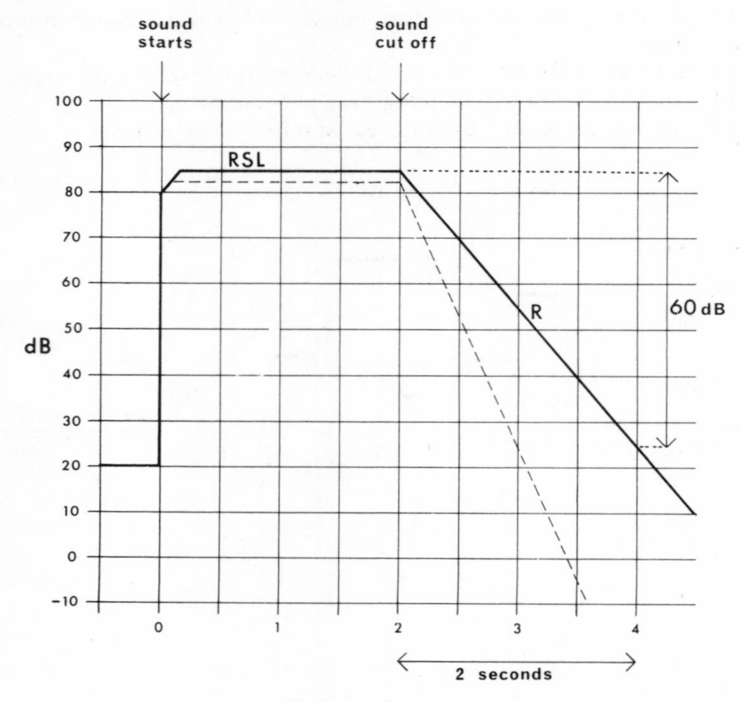

Figure 4.21

The graph also shows, in the case of a sustained sound, the initial 'build-up' of sound to a 'reverberant sound level' (RSL) before the sound is terminated at source.

The above definition of reverberation time so far omits consideration of factors (4) and (5) above. In fact criteria for 'optimum reverberation times' must assume that standing wave patterns will be prevented by good design, that is, the exclusion of parallel opposing and reflecting surfaces. On the other hand, the varying sensitivity of the ear at different frequencies is taken into account by establishing criteria for optimum reverberation times at low, middle and high frequencies.

Figure 4.22 shows recommended reverberation times for a number of functions, including speech. It also introduces a further factor, a psychological one, in relating optimum reverberation time to room volume. By general experience people expect reverberation to be longer in a large room than in a small one. Criteria are based on surveys of opinion and it has been found that the size of the room affects people's judgements of what is ideal for various purposes. The graph, figure 4.22, takes account of this.

For good speech intelligibility reverberation has therefore to be controlled by the limitation of volume or the introduction of absorbent materials, or both. The resulting reverberation time will depend on the inter-play of these two factors,

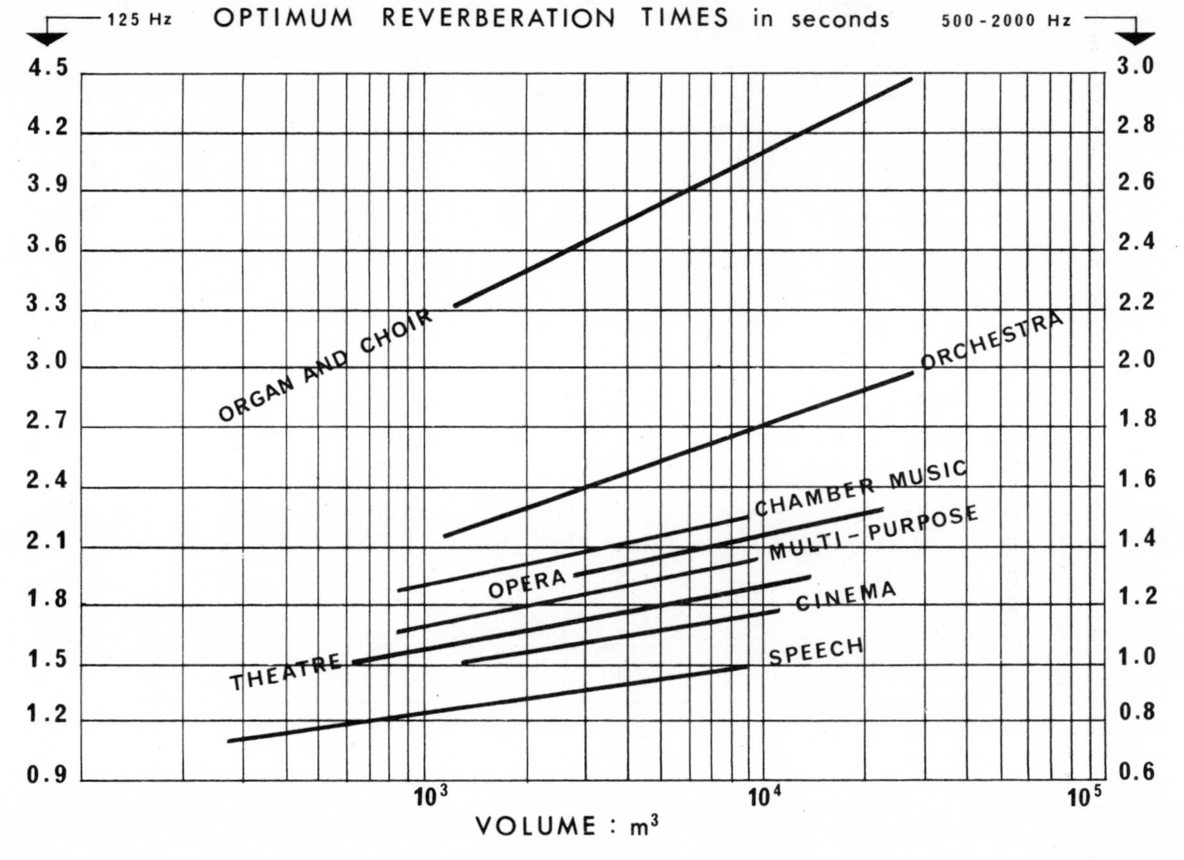

125 Hz
OPTIMUM REVERBERATION TIMES in seconds
500 - 2000 Hz
4.5
4.2
3.9
3.6
3.3
3.0
2.7
2.4
2.1
1.8
1.5
1.2
0.9
3.0
2.8
2.6
2.4
2.2
2.0
1.8
1.6
1.4
1.2
1.0
0.8
0.6
ORGAN AND CHOIR
ORCHESTRA
CHAMBER MUSIC
MULTI - PURPOSE
OPERA
CINEMA
SPEECH
THEATRE
10^3
10^4
10^5
VOLUME : m^3

Figure 4.22

as expressed by the Sabine formula for its calculation

$$RT = \frac{V}{A} \times 0.1608^*$$

where V is the volume in cubic metres and A the total absorption of the room in square metre sabins (see section 1, p. 35).

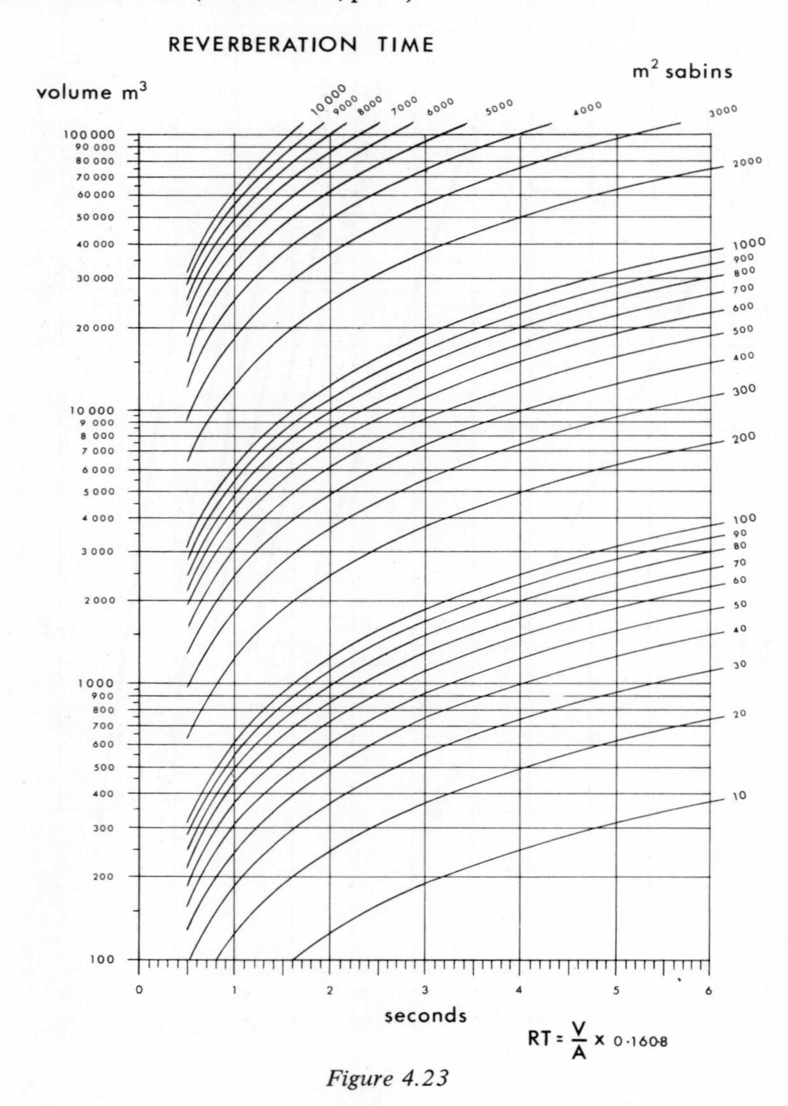

Figure 4.23

*If imperial measurement is used the constant is 0.049.

Figure 4.23 provides a graph from which can be read RT when V and A have been calculated. Alternatively, where volume has been established, the required total absorption can be ascertained for a desired period of reverberation.

In designing a room for speech, economy of seating area (and therefore floor area), together with the provision of a reasonably low ceiling to act as a reflector, will usually result in a sufficient limitation of volume. The prevention of echoes by absorbent treatment, together with the sound absorption of the audience, will often provide the degree of absorption required. A typical exception to this is the council chamber where at times the number of people present is small in relation to volume.

It remains then to calculate reverberation time on the basis of the initial design and to make any necessary adjustments to volume and/or surface materials that may be necessary.

It should also be noted that this calculation has to be made in most cases in respect of a varying occupancy since the audience is usually the most significant factor of absorption. The designer will be required to make a common sense value judgement in relating his criterion for RT to audience size where this is subject to variation. For speech acoustics it seems sensible to relate the criterion to the almost minimum likely audience since speech clarity is of pre-eminent importance. The provision of seats designed to absorb sound can, however, partly compensate for a variation of audience size. An example calculation is given on p. 194.

Duplication of sound source by loudspeakers

As stated previously, the acoustics of a room must be considered for all possible positions of sound source. The introduction of loudspeakers will add additional sources of sound of a potentially powerful nature which can result, at best, in a lack of realism and, at worst, in disturbing echoes.

The introduction of even one loudspeaker in an auditorium means that members of the audience will hear two sounds arriving at different times and from different directions. Herein lies the first problem. Figure 4.24 shows two possible arrangements of loudspeakers, one centrally placed above the platform and the other with a loudspeaker at each side. In both cases the time intervals, as found by comparing the lengths of the sound paths shown, should be checked to ensure that they are not more than about 1/30 s. In the case of a centrally placed loudspeaker (or cluster) over the platform it is possible to avoid an undue delay by keeping it reasonably low. In a wide auditorium, however, the placing of loudspeakers at each side can cause an excessive delay.

The second problem arises when a loudspeaker is introduced, say, half way down a large auditorium for the benefit of people at the rear, as shown in figure 4.25. This has the advantage that the power of the loudspeaker can be reduced (as compared with a loudspeaker near the platform) but can result in a time interval between the sound emanating from the human speaker and that from the loudspeaker—for a different reason. Figure 4.25A explains this. The sound reaching the listener (L) from the speaker (S) travels at 340 m/s through the air. The sound emanating from the loudspeaker much closer to the listener has travelled along a wire at the speed of light. He will hear the loudspeaker first

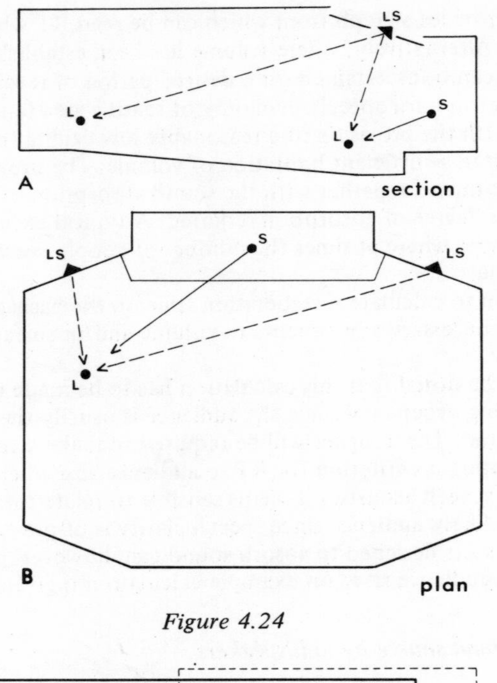

Figure 4.24

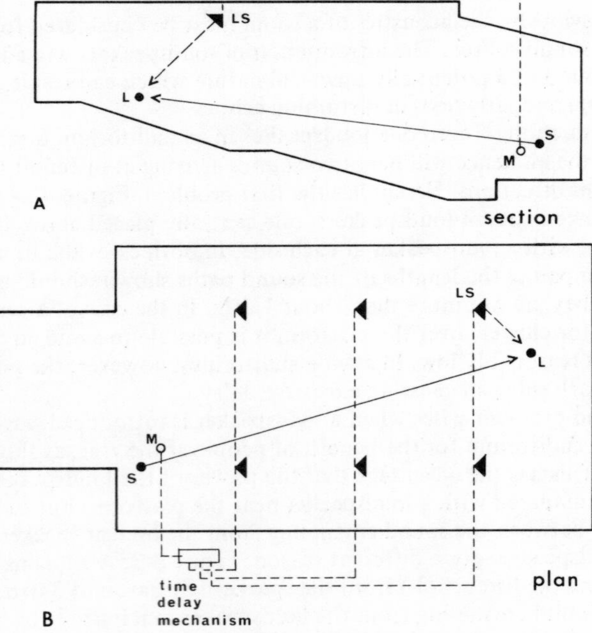

Figure 4.25

and the sound from the platform either as a weak echo or a blurring of the speech sounds, depending on the time interval.

This problem can be overcome by introducing a delay mechanism into the loudspeaker circuit so that the sound from the loudspeaker is produced a few milliseconds after the arrival of the sound through the air.

Another reason for delaying the output from the loudspeaker is that one of the ways by which we identify the direction from which a sound is coming in such a situation is by the sound which arrives first. The reader will have noticed that in cathedrals and large churches loudspeakers are introduced at intervals down the length of the nave. It is normal practice to include in the circuit a delay mechanism which varies the delay progressively for each loudspeaker (or opposite pair of loudspeakers) in accordance with its distance from the pulpit and lectern. In this way, provided the sound from the loudspeaker is not too loud, the listener will identify the source of sound as being from the pulpit rather than from his 'local' loudspeaker. The arrangement is shown diagrammatically in figure 4.25B.

It will also have been observed that 'column loudspeakers', that is, a vertically placed cluster, are invariably employed in large auditoria. These have the effect, if tilted downwards, of spreading the sound horizontally over the audience. By this means a third problem is solved, namely, the possible increase of reverberation due to the increased power of the sound. Criteria of optimum reverberation time for speech are relative to the average power of unassisted speech. There is always a tendency to operate loudspeakers at an unnecessarily high level and therefore increase reverberation. By 'fanning' most of the amplified sound over the highly absorbent audience this effect can be reduced. In interiors which have not been correctly designed for speech, such as old churches, this becomes even more important.

Apart from this, no attempt will be made to discuss the choice of equipment. This is a highly specialised field which is developing rapidly and the architect is advised to seek specialist advice at an early stage in design. The choice of the most suitable microphones, amplifiers, output delay equipment and loudspeakers is as important as the correct design of the interior. As an example: the wrong choice of microphone in relation to the position of loudspeakers can cause the phenomenon known as 'feed-back' which produces a high-pitched squeal.

Ambient noise

The term 'ambient noise' is here used to describe sound incidental to the occupation of the room but extraneous to the room's essential purpose. Such noise is mainly created by the audience itself: the shuffling of feet, the movement of tip-up seats or writing tablets, the closing of doors, etc. Some of the sound produced by the occupants of the room can only be controlled by the audience itself but the designer can reduce it by, for example, providing quiet floor finishes, rubber stops on tip-up seats and writing tablets, door closures, and so on. Apart from the stage, flooring should be solid since panel resonance will amplify footfalls. Finally, the control of reverberation will considerably reduce the nuisance caused by ambient noise.

Intrusive noise

As with ambient noise, intrusive noise can mask speech sounds and reduce clarity to an even greater extent. 'Intrusive noise' is that which penetrates the room's defences from outside and its reduction has been fully discussed in section 3. Because the ambient noise level in a well-designed auditorium with a quiet audience is fairly low, the criterion for acceptable intrusive noise level is demanding, as will be seen by reference to the table on p. 112.

Intrusive noise is not only due to road and air traffic. Some of the sources of noise external to the auditorium are

(1) foyers, corridors, lobbies, staircases
(2) restaurants, bars and lounges
(3) theatre workshops, stores and plant rooms
(4) green rooms
(5) projection rooms
(6) rain on light roof structures.

Acoustics for Music

As with speech, it is necessary first to examine the nature and power of the source of sound. This may range from a single instrument in a music practice room to a full orchestra of around 80 players plus a choir of 100 or more in a large concert hall. On the other hand we may have to consider the special requirements of church music with organ and choir. What we have called the 'nature of the source of sound' is not only a matter of size but also of character, linked with tradition. For example, it is generally accepted that the music of Mozart is best heard in an environment with a rather shorter reverberation time than is appropriate for Wagner. Handel's choral music, written for the church, ideally requires a still longer period of reverberation.

In the case of orchestral music we are concerned with sounds which can vary from 30 dB.A or less in quiet passages, to 80 dB.A or more, as measured in the middle of the auditorium. This range of sounds must be heard without distortion and, in particular, the weaker components in music—harmonics and transients*—require to be heard in seats at the rear of the hall.

Furthermore, the range of frequencies which are important for the full appreciation of music is much greater than in the case of speech. The development of high-fidelity sound reproduction has underlined the importance of harmonic frequencies well above the highest note played by the orchestra. Another important difference between speech and orchestral music is the spread of the sound source. It may extend over a width of 20 m and a depth of 10 m if we include orchestra and choir.

But perhaps the most striking difference between speech and music, and one which makes an objective appraisal of requirements difficult, is in the far more

*Transients are weak sounds of very short duration produced by most instruments in advance of the sustained note, for example, the first touch of the bow on the strings of a violin. By transients and harmonics we identify instruments.

complex aesthetic demands made by the audience. This has given rise to a whole vocabulary of expressions such as 'intimacy—warmth—texture' which the designer has to try to interpret in objective acoustic terms. In what follows such concepts will be related to the design of medium-to-large concert halls. Notes on other types of auditoria for music will be found under 'Auditorium types' on pp. 178 to 191.

As with speech, we can begin our consideration of the acoustic requirements for orchestral music under the broad headings of 'power' and 'clarity'. To these however must be added two other factors: 'blend' and 'ensemble', the first because we are dealing with a multiple sound source, the second because the final aesthetic result is dependent on the mutual co-operation of many individuals. These four aspects can then be considered in relation to their components as set out below. Under these headings most of the acoustic design requirements for concert halls should emerge.

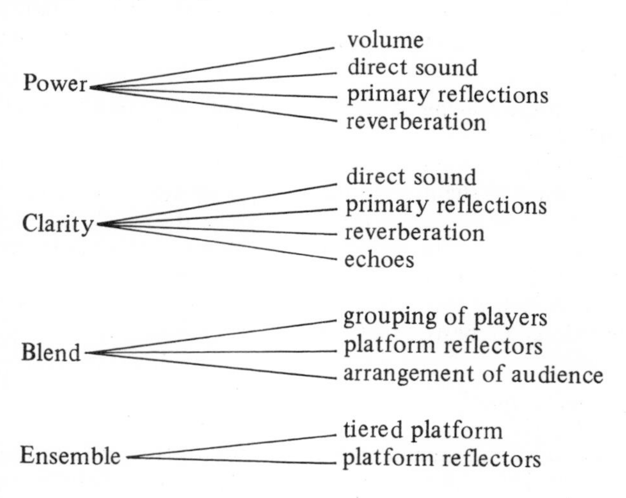

Power and volume

A small orchestra playing in a large concert hall is likely to produce a body of sound which is too weak to be heard satisfactorily above the level of audience noise. The 'acoustic density' of reverberant sound may be inadequate. This is quite apart from the fact that the distance from platform to rear seats may be too great for the adequate reception of direct sound. On the other hand a large orchestra in a too-small hall can produce an uncomfortably high level of sound. A fundamental requirement in the design of a concert hall will therefore be a suitable relationship between platform accommodation and seating capacity. Since some variation in the size of orchestras will be unavoidable, consideration should be given to reducing the volume of the platform space by movable screens, if not to the possibility of changing the volume of the auditorium. In halls where choirs are accommodated behind the orchestra provision should certainly be made for screening this area when not required.

Power and direct sound

The reduction in the power of direct sound towards the rear of the auditorium will be due to distance, but also possibly due to audience absorption or the mutual obstruction of sound between the players. It should be noted that although the attenuation of sound due to distance in the case of a full orchestra is less than for a 'point source', the attenuation of sound from solo instruments follows the inverse square law. Therefore the need to reduce distance from platform to rear seats for the weaker sounds of solo instruments is as important as in the case of speech.

Audience absorption of direct sound, as described on p. 147 may again add further attenuation, in particular affecting transients and harmonics by which we appreciate the distinctive qualities of each type of instrument. If players at the front of the platform are so arranged that they form an obstruction to the sound from players behind them, there will be an additional weakening of sound by scattering and absorption. Therefore, to maximise the power of direct sound at the rear of the hall

(1) economise in the spacing of seats and the total area of gangways within the body of seating

(2) subject to the provision of strong lateral reflections, as described below, reduce the distance to the rear of the hall by a square rather than elongated proportion

(3) in large halls consider the inclusion of a gallery or galleries to bring rear seats nearer the platform, provided the overhang is shallow

(4) reduce the audience absorption of direct sound by well-raked seating

(5) provide a tiered orchestral platform.

Power and primary reflections

Reflecting surfaces around and immediately in advance of the orchestral platform can serve two purposes. They can reinforce direct sound—most usefully towards the rear of a large hall—and they can give rise to a sense of enclosure or 'intimacy' which seems to be a characteristic of the most popular concert halls.

The use of reflectors for the first of these two purposes has been fully discussed under the heading 'Acoustics for speech'. It is, however, necessary to stress that in the case of orchestral music the over-employment of reflectors can result in so much sound being directed towards the highly absorbent audience and rear walls that reverberation is much less than calculated, and inadequate. Plain reflectors should, at the most, be limited to surfaces around and above the orchestra.

Flank wall reflectors, which may be modelled, serve to provide strong lateral reflections to supplement direct sound provided the width of the hall near the platform is not too great. Such reflections should arrive not later than 1/30 s after the reception of the direct sound, in relation to a position in the middle of the seating area. It is generally considered that the first ceiling reflection should arrive after the lateral reflections. Such an arrangement, while providing what musicians call 'intimacy' also ensures the adequate generation of reverberation.

This means that, although the fan-shaped auditorium has the advantage of placing more of the audience at a comfortable acoustic and visual distance from the orchestra, it should not depart too far from the traditional rectangular plan of limited width.

The reader will have observed that in some recent examples of concert halls small independent suspended reflectors (sometimes called 'acoustic clouds') cover the ceiling area. The report that in one such hall the sound is rather too 'brilliant' may be due to the fact that small reflectors reinforce the higher frequencies but, by edge diffraction, the middle and low frequencies are absorbed in the space behind.

Since the more powerful and directional instruments are normally placed towards the rear of the platform, it seems logical to place the angled overhead reflector above the front of the stage where it can also assist soloists. The remainder of the platform ceiling can be considered as an aid to maintaining 'ensemble' (see below). In both cases, to be effective the height should be limited to between 5 and 10 m from average platform level. It will already be seen that the design of a concert hall begins with the design of the orchestral platform and possibly the accommodation of a choir and organ.

Power and reverberation

Reverberation can increase power and provide the 'body' or fullness of tone which has come to be expected of orchestral music. The dynamic range of the players is increased and by a build-up of reverberation effective crescendos can be achieved. On the other hand excessive reverberation can reduce clarity beyond what is acceptable, depending on the type of music. Figure 4.22 suggested appropriate reverberation times in relation to broad categories of music and auditorium volume. The relationship between the reverberation times suggested and the acoustic environments in which various types of music developed will be evident. The variation with room volume is partly because less reverberation is expected in smaller rooms and partly because room size will be reflected in the size of the orchestra or group, and therefore the type of music.

It will also be seen that longer periods of reverberation are suggested at low frequencies than at middle and high frequencies. This is because, as will be seen by reference to figure 2.3, the ear is much less sensitive to low frequencies. If adequate low-frequency reverberation is not provided such instruments as cellos will lack adequate power in the bass and, in general, the auditorium will be said to lack 'warmth'. This means that in selecting internal finishes care has to be taken to limit the use of those which absorb sound preferentially at low frequencies. If therefore wood panelling is used the air space between the fixing battens will probably have to be filled solid to reduce low-frequency resonance.

It is equally important to provide adequate reverberation at middle and high frequencies and, since the audience will contribute considerable absorption in this part of the frequency range, any additional absorption by porous materials has to be limited. A hall lacking in middle and high frequency reverberation is said to lack 'liveness' and, if very high frequencies are weak, to lack 'brilliance'.

The quality of reverberation, as well as its duration, has to be considered. After the reception of direct sound and first reflections from surfaces near the orchestra, reverberation should decay steadily and without distortions due to standing waves. Reverberation should also be 'diffuse', that is, as nearly as possible equal in all parts of the auditorium.

The placing of reflectors around the platform, together with the bold modelling of all other surfaces will contribute to a steady and diffuse reverberation. The avoidance of parallel opposing and reflecting surfaces will reduce the risk of standing waves. On the other hand, the localisation of absorbent material will adversely affect diffusion by lowering the reverberation time in these areas. An example of this is the somewhat 'dead' nature of the acoustics under deep galleries where the relationship between absorption (audience and rear wall) and volume is high, as shown in figure 4.26. The opposite will be the case at the front of the hall. To some extent this difference is reduced subjectively because direct sound is more powerful near the platform.

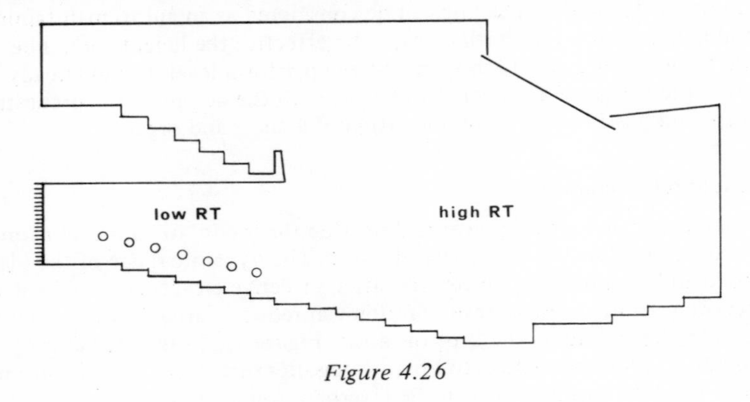

Figure 4.26

Clarity and direct sound

The clarity of rapidly articulated sounds against a background of reverberation will in part depend on the limitation of reverberation. It will also however depend on the preservation of the power of such sounds by limiting the distance to rear rows of seats, by reducing audience absorption of direct sound and by providing uninterrupted sound paths from a tiered orchestral platform.

The clarity of the higher frequencies is often referred to as 'brilliance', which can be obtained, not only by the above means but also by well-placed reflectors with a low coefficient of absorption at high frequencies.

However, since audience absorption of direct sound at grazing incidence is greater at high frequencies than low, the raking of seats is the first step in preserving brilliance and the clarity of the more distinctly articulated sounds in music.

Clarity and echoes

The greatest loss of clarity will of course result in the presence of echoes or 'near-echoes', the prevention of which has been fully discussed on pp. 156 to 159. It is, however, appropriate to mention here one of the difficulties experienced in concert hall design in providing sufficient reverberation, especially under galleries. This is exacerbated if the rear wall has a concave curvature requiring a high coefficient of absorption to reduce concentrated echoes. The provision of a convex curve for the rear wall, although perhaps unnatural in relation to normal seating arrangements, would reduce the need for a high coefficient of absorption for the rear wall, especially if the surface were also boldly modelled to disperse reflected sound. As far as the author knows, this solution has not been tried in practice but the arrangement is shown in figure 4.27B compared with the more usual plan in figure 4.27A.

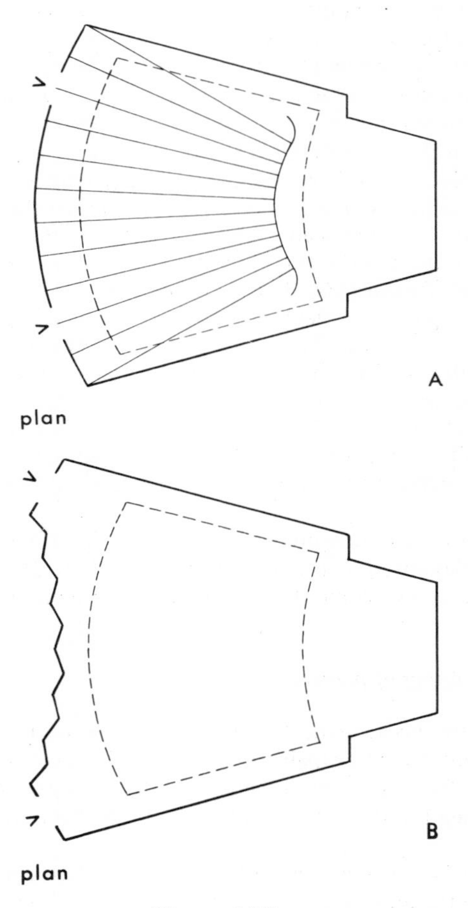

Figure 4.27

An example of a particularly serious form of echo was shown in figure 4.18, which can occur in opera houses in relation to the orchestral music or the singing. Because of the height of the interior and the presence of a shallow dome a concentrated echo can occur as shown by the large difference in sound path between direct and reflected sound. Moreover, in this case the echo may be louder than the initial sound and will sweep across the first gallery as singers move from one side of the stage to the other. Such a ceiling, even without the dome, would require sound-absorbing or dispersive treatment.

Clarity and reverberation

It will have been appreciated that the need to provide sufficient reverberation for 'body' and tonal quality for some kinds of music is in conflict with the need for clarity, especially for rapidly articulated sounds such as a cadenza on the piano. The criteria for optimum reverberation times given in figure 4.22 effect a compromise which is generally found satisfactory. The conflict is partially resolved by the fact that such articulated sounds from solo instruments are weaker than full chords or combinations of instruments and do not generate so much reverberation. If, however, such passages of music are accompanied by the full orchestra, excessive reverberation may seriously reduce clarity.

Reverberation will of course also vary with the size of the audience, although most concerts in this country seem to be well attended. The ultimate answer to both problems, variation in the type of music or the size of the audience, would be in the provision of panels distributed around the body of the hall which can be reversed, exposing either reflecting or absorbing surfaces as the need arises. These would be motor-operated and controlled from the platform. Such a device, but manually operated, has been employed on a smaller scale.

However, in the case of recent concert halls, the problem seems to have been the provision of *sufficient* reverberation. This resulted in remedial action at the Royal Festival Hall by means of electronically assisted reverberation. In this way, reverberation was increased for the lower frequencies, adding 'warmth' to the musical tone.

The difficulty of accurately predicting reverberation time at the design stage of large and complex interiors is examined on pp. 193 to 197. A scale model technique which may overcome the difficulty is described on p. 197.

Blend and the grouping of players

The composer writes his score on the assumption that his music will be heard as a blend of instrumental sound capable of a far wider range of expression than could be achieved by a single instrument. It is therefore the duty of the designer to preserve this blend in the design of the concert hall and in particular of the platform.

If players were, for example, spread out across a platform the full width of a large concert hall, the audience, especially near the front, would be conscious of the fact that the sound of the violins had been separated from that of the cellos and the blend of the two groups would be diminished. The normal close-grouping of players around the conductor and the limiting of spread by a fairly deep seating arrangement should therefore be maintained in designing the platform.

Blend and platform reflectors

Vertical reflectors behind the players also contribute to the blending of the total sound, as well as providing additional power when needed. Figure 4.28A shows that the sound from instrument b will be blended, by reflection, with the sound from instrument a. The general effect of such reflectors is therefore a blending of the total sound of the orchestra and its projection towards the audience. This is largely absent in the example shown in figure 4.28B.

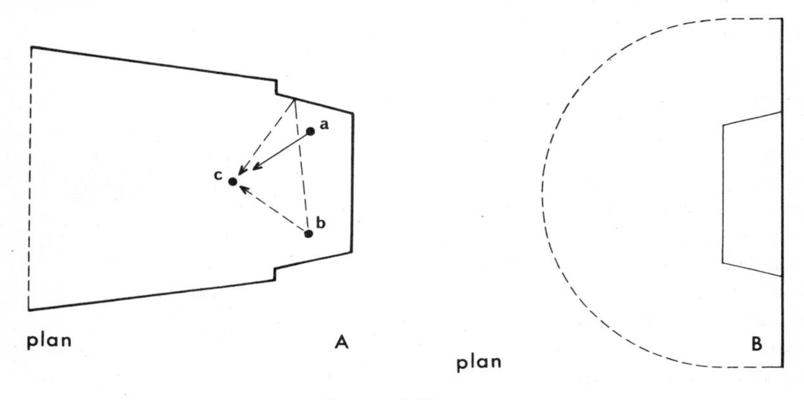

Figure 4.28

This blending of instrumental sound appears at first glance to run counter to the aims of those who make stereophonic gramophone records, from which the listener can distinguish the left and right in the arrangement of players. The purpose of stereophonic sound reproduction is in fact to give an air of realism—of 'feeling that you are there'—part of which is sensing the spread of the orchestra and first reflections from the walls of the hall. This experience will remain however closely the players are grouped.

While on the subject of the orchestral platform, it should be mentioned that this should be boarded, with an air space below, in order to provide 'panel resonance'—in other words, staging and not wood flooring direct on concrete. This is particularly important for instruments in contact with the stage, such as cellos.

Blend and arrangement of audience

The conductor at rehearsals is, among other things, concerned with very subtle modifications of the power of the various instruments in order to achieve what he considers to be the best interpretation of the score. He does this from a central position.

It follows therefore that only that part of the audience more or less centrally disposed behind him will hear the music as he intends. Although this must not be interpreted too literally, even if it were practical to do so, it is questionable whether many seats should be disposed at the sides of the platform, as would be the case in the plan shape in figure 4.28B. The plan shape in figure 4.28A, on the other hand, disposes most of the seats in a position where 'balance' is preserved.

In any case, plan B presents two further problems

(1) the rear, potentially echo-producing, wall would require an extended absorbent treatment which would make the provision of an adequate reverberation time difficult and

(2) there remains only the ceiling which can be effectively employed to scatter sound and provide a diffuse reverberation.

Ensemble and platform reflectors

To make it possible for the instrumentalists in an orchestra to keep perfectly in time with each other and adjust their playing to the conductor's intentions, it is necessary for them to hear each other. They also desire to sense the power of their playing as a group.

The best way to achieve this is to provide a reflector above the players which will project the sound of each instrument to all the other players in the group in addition to the sometimes interrupted direct sound.

Such a reflector is shown in figure 4.29 and it should be noted that it is placed fairly low and not parallel to the horizontal plane of the platform, to avoid the risk of standing wave patterns.

Summary

Figure 4.29 attempts to demonstrate all the above design principles in terms of a possible plan and section for a concert hall. It must, however, be stressed that this is not intended to suggest that there is only one satisfactory design; many alternatives are possible when the principles are understood.

Finally, reference should be made to what has been stated under the headings of 'Ambient and intrusive noise' on p. 167, which applies equally to the design of auditoria for music.

Multi-purpose Rooms

Many auditoria are used for both speech and music, examples being churches, town halls, school assembly halls and variety theatres. As we have seen, the acoustic requirements for speech and music conflict to some extent, mainly in respect of optimum reverberation time and the employment of reflectors. Some halls are required for a great variety of other activities such as dancing, physical training and exhibitions. How then can these varied activities be provided for with acceptable acoustic conditions?

There are perhaps three approaches to the problem

(1) to properly satisfy one requirement, if it is found that for most of the time one activity will predominate

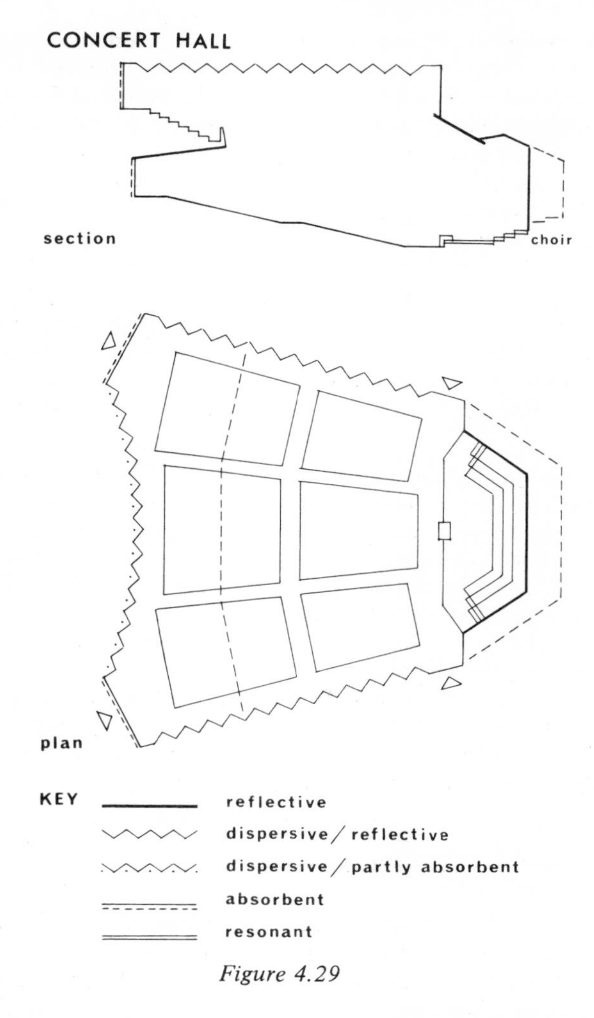

Figure 4.29

(2) to compromise, where more than one activity requires equal consideration, or

(3) to so design the auditorium that its acoustic properties can be changed to suit different functions.

If we consider approach (1) first, then the designer must by analysis and calculation satisfy himself that acoustic conditions will not be completely unacceptable for the remaining activities. In this regard, he must remember that the number of people engaged in other activities may be very different from the one he deems to be the more important. As has been explained, variations in occupancy can make a considerable difference to the degree of reverberation which results.

Approach (2) is likely to involve a compromise in the period of reverberation chosen, as is indicated in figure 4.22 for multi-purpose rooms. It may also mean that; where a level floor is required for dancing or other functions, there will be some sacrifice of good speech acoustics due to audience absorption of direct sound (not to mention unsatisfactory sight lines). This disadvantage can in some cases be minimised if only the area required for dancing is level and the remaining area is tiered. It should also be remembered that, where the audience is seated on a level floor, the stage should be as high as possible and that low, well placed reflectors should compensate for the attenuation of sound passing over the heads of the audience at a small angle of incidence.

It is, however, only possible to make reasonable compromises if the optimum requirements for the various functions have each been carefully studied.

The third approach (3) opens up the possibility of achieving far better results all round but some of the devices which will be mentioned are expensive and some require knowledgeable operation or maintenance which is not always available. Some possibilities are listed below but the designer will no doubt discover others.

(1) A reduction in volume of the hall by curtains or sliding screens will reduce reverberation when occupancy is reduced. An example of this is the closing off of a balcony volume along the line of the balcony apron.

(2) Alternatively, changing the absorption characteristics of vertical surfaces is possible either by drawing curtains or providing rotating panels which have one side reflecting and the other absorbent.

(3) The mechanical tilting of a level floor when required will provide better sight lines for a seated audience and reduce the absorption of direct sound.

(4) the use of retractable tiered seating over part or all of an otherwise level floor is an alternative to the above which has frequently been adopted.

(5) It is now possible to provide electronically assisted reverberation for music in an acoustic environment designed for speech.

Auditorium Types: Summary of Requirements

In what follows, brief mention only is made of the main acoustic considerations, as a form of check list. Earlier parts of this section should be consulted for a more detailed explanation of most of the factors and reference is made to the relevant pages, where applicable.

The examples are in alphabetical order for ease of reference and each check list is complete in itself.

Bandstands

(1) Where these take the form of 'shells' the design is similar to open-air concert platforms, to which reference should be made below. p. 187

(2) Where the audience is seated around a central bandstand, the soffit of the roof forms the main reflector and should be saucer shaped, presenting its convex surface to the band. p. 148

(3) Sliding glass screens protect the band from the wind and act as useful reflectors.

(4) The stage and apron should be designed as resonators. p. 175

(5) A sloping or terraced seating area around the bandstand will improve the reception of direct sound. p. 147
(6) If possible, the audience should be protected from wind and external noise by screen walls or planting. p. 89

Churches

(1) The plan and seating arrangement of a church may be much more determined by liturgical considerations than acoustics.
(2) Apart from this, acoustic design is made difficult by the conflicting requirements of speech and music, especially in the case of choral and organ music which require a long period of reverberation. p. 163
(3) The kind of compromise adopted will depend upon the relative importance of speech and music in each case. p. 176
(4) Considerable variation in the size of the congregation may usually be expected and this will also influence the choice of a compromise reverberation time.
(5) The height usually required for aesthetic reasons provides a volume per person which makes control of reverberation for speech unusually difficult.
(6) A possible approach for meeting the different requirements for speech and music is to provide a reverberant volume for the organ and choir while keeping reverberation as short as possible in the body of the church. The position of the pulpit and lectern in advance of the chancel makes this possible in Anglican churches.
(7) If this is done, calculation of reverberation times in the two volumes becomes very approximate, even when they are clearly defined by a chancel arch. With an open plan, only an average period of reverberation can be calculated as a rough guide.
(8) Hard reflecting materials around the choir (and organ) provide the necessary reverberant conditions.
(9) In the body of the church the most convenient position for absorbent materials is likely to be the ceiling or roof lining. Where the section is vaulted or pitched in form, absorbent materials here will reduce the localised reverberation associated with these forms.
(10) Apart from this, absorbent materials may with advantage be distributed around wall surfaces in the body of the church if the control of reverberation for speech requires it.
(11) The wall opposite the pulpit and lectern (or their associated loudspeakers) may require treatment to reduce echoes. p. 156
(12) Where chapels form separate volumes, these may become independent reverberant chambers if their acoustic treatment is not balanced with that of the church.
(13) A low, angled reflector over the pulpit (the traditional 'sounding board') is essential and, if the pulpit can be placed in a re-entrant angle, so much the better. p. 148
(14) In large churches, electronic amplification of speech may be necessary.

Cinemas: monophonic sound

(1) The plan shape of a cinema is influenced by sight lines to an even greater extent than the theatre. Because of the higher average level of sound, this can take precedence over acoustic requirements in determining the shape of the seating area.
(2) The introduction of a balcony reduces the distance from the screen of the furthest seats in a large cinema and can reduce the number of seats uncomfortably close to the screen. The directional nature of the loudspeaker allows balconies to be deeper than in a theatre.
(3) Although the sound may be raised to any level to reach rear seats, the result may be unsatisfactory in the front seats of a large cinema. In these cases overhead reflectors can well be used to provide progressive reinforcement of sound towards the rear. p. 153
(4) For the design of reflectors, the intended height of the loudspeaker must first be established. It should normally be centrally behind the screen on plan.
(5) Side walls should be dispersive, with areas of absorbent material as required to reduce reverberation.
(6) The rear wall must be absorbent and dispersive. p. 156
(7) The rake of the floor may be less than in a theatre, provided clear sight lines are available for all members of the audience to the bottom of the screen and the loudspeaker is not too low. p. 147
(8) Check for echoes from all re-entrant angles. p. 157
(9) The surfaces of any volume behind the screen are best made absorbent.
(10) Allowance should be made for the usually variable size of audience and seats should be as sound-absorbing as possible. p. 202
(11) Reverberation time* should not be more than about 1 s because reverberation is added to the sound track when required. p. 163
(12) In designing a cinema for monophonic sound, the possibility of later introduction of stereophonic sound equipment should be taken into account–see below.

Cinemas: stereophonic sound

(1) The introduction of stereophonic sound changes the whole approach to the design of the cinema since loudspeakers are distributed all around the auditorium.
(2) It is essential that the 'directional effect' is not confused by lateral reflections. This means that all wall surfaces, including the screen surround, should be absorbent or partially absorbent and dispersive.
(3) A 'directional form' for the ceiling is also inappropriate and a more or less horizontal and dispersive surface is advisable.

*In this section approximate reverberation times are given for middle and high frequencies, but see figure 4.22.

(4) Prevention of echoes will follow from the above treatment but the design must be checked for echo-producing corners and reflections from balcony aprons, if any. p. 157
(5) Allowance should be made for the variable size of audiences and seats should be as absorbent as possible. p. 202
(6) If possible the interior should be designed so that loudspeakers at the sides and rear need not be too close to nearby seating.
(7) Reverberation time* should not be more than about 1 s because reverberation is added to the sound track as required. p. 163

Classrooms

(1) Though small, a classroom may be much too reverberant if all surfaces are reflecting. This is unsuitable for speech and creates a noisy environment.
(2) The ceiling, rear wall and wall opposite the main windows should be sufficiently absorbent to control reverberation.
(3) Perforated acoustic tiles or slotted fibreboard are suitable for ceilings with soft pinning boards on the walls.
(4) Reverberation time should be not more than 3/4 s. p. 163

Committee rooms

(1) Seating arrangements will vary according to the size of the room but the aim should be to seat members facing each other in a circular or oval formation. Long parallel ranks of seats are undesirable.
(2) The volume of the room and the height of the ceiling should not be allowed to become too great in an attempt to create a spacious effect.
(3) The volume of the room in relation to the number of persons present usually requires the employment of absorbent material to control reverberation. It is best applied to wall surfaces, especially in rooms of large plan dimensions. p. 163
(4) In large committee rooms echoes, or 'near-echoes' can be avoided by the absorbent treatment of wall surfaces. p. 150
(5) Reflecting re-entrant angles can also give rise to echoes. p. 157
(6) A low horizontal ceiling, or a lowered portion over the committee table, provides reinforcement of sound for all positions of source, in the case of large rooms. p. 152
(7) A high ceiling causes delayed reflections and is better made dispersive.
(8) Provide well-upholstered seating and soft floor finishes.
(9) The reverberation time should not be more than about 3/4 s. p. 163

*In this section approximate reverberation times are given for middle and high frequencies, but see figure 4.22.

Community centre halls

(1) Ideal acoustics are normally unattainable in multi-purpose halls for the following reasons
  (i) the variety of uses, requiring different periods of reverberation
  (ii) the considerable variation in the occupancy of the room
  (iii) the need to provide a level floor for dancing. p. 176
(2) Subject to the requirements of sight lines for the stage, reduce the length of the room in proportion to its width as much as is practicable.
(3) Provide tiers for the rear seating if possible. p. 147
(4) A shallow balcony will also provide seating with good sight lines and acoustics, and reduce the number of seats required on the level floor. p. 147
(5) Employ an angled reflector over the stage to reinforce sound at the rear seats and to compensate for inferior direct sound. p. 153
(6) Provide an apron stage for dramatics and musical entertainment, to increase the value of the reflector.
(7) The stage should act as a resonator. p. 175
(8) Padded, preferably fabric-covered, nesting chairs are preferable to wood seats. p. 202
(9) If possible, arrange for part of the hall to be curtained off at times of small occupancy. p. 176
(10) Side walls should preferably be dispersive. Window reveals and projecting piers help to meet this requirement.
(11) If the room is likely to be used frequently by small numbers of people, it may be necessary to make the ceiling absorbent, or at least dispersive, to reduce strong inter-reflections between exposed floor and ceiling, confining the reflector to the part of the ceiling near the stage.
(12) The rear wall should be sound-absorbing, especially in rooms of long proportion. p. 156
(13) Check for possible echoes from the re-entrant angle between rear wall and ceiling. p. 157
(14) In most cases, speech is the more important function and its intelligibility is more critical. A short period of reverberation is also appropriate for films, dancing and at times of small occupancy. A reverberation time of about 1.25 s for the average size of audience expected is a reasonable compromise between the requirements of speech and music. p. 163

Concert halls

(1) Various plan shapes can be satisfactory acoustically provided the plan proportion is not too long and the following conditions are met.
(2) A rectangular plan has the advantage of providing the strong lateral reflections which generate reverberation and provide 'acoustic intimacy'. p. 170
(3) For large concert halls the fan-shaped plan has the advantage of reducing the distance to rear rows of seats and placing more of the audience at an optimum distance from the platform. p. 177

(4) If a fan shape is adopted however the angle of flank walls must be limited so that strong lateral reflections with a short delay time are retained (see below). p. 170

(5) Alternatively, the distance from platform to rear rows of seats may be reduced by the introduction of a gallery, provided it is shallow. p. 172

(6) Seating should be well raked to provide strong direct sound. p. 147

(7) Provide a reflecting surface behind the orchestra and angled reflectors at the sides. p. 175

(8) A limited reflector over the front of the platform may be provided to assist soloists but otherwise this surface should reflect sound back towards the players. Reflectors over the platform must not be parallel to its surface. p. 176

(9) Flank walls should be reflective and dispersive to provide lateral primary reflections with a short time lapse—not more than 1/30 s in relation to a position in the centre of the seating area. p. 170

(10) The ceiling surface should be dispersive and at a height which will provide reflections which arrive after the flank wall reflections.

(11) The rear wall (and gallery apron, if any) should be dispersive and absorbent to prevent echoes. If these surfaces are concave a greater degree of absorption will be necessary. p. 172

(12) Check carefully for possible echoes from re-entrant angles. p. 157

(13) The over-all aim is to provide strong initial sound by direct path and closely following primary reflections, with a background of diffuse and steady reverberation.

(14) Reversible absorbent/reflecting panels can be used to modify reverberation for various types of music, variations in the size of audience and for 'tuning in' the hall after completion.

(15) Sound-absorbing seating and quiet floor finishes are essential.

(16) The optimum reverberation time at middle and high frequencies varies between 1.5 and 2.25 s, depending on the size of the hall. At lower frequencies it should be longer. p. 163

Conference halls

(1) In the usual arrangement of seating, where the body of the assembly faces the platform, speakers 'from the floor' have their backs turned to part of their audience. A wide rather than deep plan-form reduces this disadvantage. p. 192

(2) A curved arrangement of seating in relation to the platform encourages speakers to turn towards the majority of their audience. p. 192

(3) If the table for those seated on the platform is also curved in the opposite direction, they can see and hear each other more easily. p. 192

(4) The 'equal-intelligibility curve' may well be used to assess audibility in relation to extreme positions of the various sources of sound. p. 143

(5) Strict economy in seating area and volume is essential if sound amplification is to be avoided in larger halls.

(6) A raked floor provides better direct sound paths for speakers on the platform. A tiered floor is usually considered inconvenient because of the need for delegates to move freely during the proceedings. p. 147

(7) The platform can be a little higher than, say, a theatre stage, to reduce audience absorption of direct sound. p. 148
(8) A low, horizontal reflecting ceiling provides reinforcement of sound over the seating area for any position of source. p. 152
(9) An angled reflector over the platform provides reinforcement of sound at the rear of the hall and, conversely, reflects sound from the rear of the room to the occupants of the platform. p. 192
(10) Where galleries are introduced, they should be shallow and must not project over the seating area. Occupants of galleries may depend entirely on ceiling reflections when listening to speakers they cannot see. Seats should therefore not be placed under gallery soffits. p. 156
(11) All walls should be dispersive or absorbent to reduce echoes or near-echoes relative to the variety of speaking positions. Walls opposite loudspeakers (if used) should be absorbent and dispersive. p. 156
(12) Check carefully for the possibility of echoes from re-entrant angles and balcony aprons. p. 157
(13) Consider the practicability of controlling reverberation, for small attendances, by temporarily reducing the volume of the hall. p. 176
(14) Provide sound-absorbing seating.
(15) The continual movement of delegates, usual at conferences, requires the provision of fixed seating and quiet floor finishes.
(16) If electronically amplified speech is necessary, great care must be taken in the positioning of microphones and loudspeakers. p. 165
(17) Reverberation time should be in the region of 1 s, taking into account partial occupancy of the hall. p. 163

Council chambers

(1) Subject to the comfortable spacing normally required, form the seating into a compact group.
(2) The traditional semi-circular or 'horse-shoe' arrangement of seating is the best acoustically, as the councillors then face each other when speaking.
(3) Tiered seating is preferable acoustically and the chairman's table should be on a raised dais. p. 147
(4) Provide a horizontal reflecting ceiling over the seating area, as low as reasonably possible.
(5) The ceiling reflector should be extended to reflect sound into the public gallery, and the gallery should not overhang the main seating area. p. 152
(6) An angled reflector over the dais is optional but, if lower than the ceiling, will give the chairman an acoustic advantage in controlling the meeting. p. 192
(7) All walls should be dispersive or absorbent to reduce delayed cross-reflections. Absorbent material on walls is in any case likely to be required to control reverberation.
(8) The angles between walls and ceiling should be designed to prevent diagonal echoes. p. 157

(9) The gallery walls should be sound-absorbing and quiet floor finishes are essential.
(10) Well upholstered, fabric-covered seating reduces the extension of reverberation time at small attendances.
(11) The reverberation time should not be more than 1 s for a quorum. p. 163

Debating halls

(1) Seating arrangements vary considerably according to procedure and the constitution of the assembly, but the overriding consideration is economy of plan area and volume.
(2) Where members face each other, on the House of Commons pattern, rows of seats should be tiered and not longer than necessary.
(3) The extended U-shaped arrangement tends to reduce the length of the room for a given number of seats, and has this advantage over the House of Commons arrangement.
(4) The semi-circular arrangement of seats is probably the best acoustically.
(5) The ceiling should be horizontal and reflecting, and not more than 6 m above speaking level. It may with advantage be much lower than this in smaller rooms. p. 152
(6) An angled reflector over the chairman or 'speaker' will assist him in controlling the debate. p. 192
(7) Where galleries are introduced they should be shallow and must not project over the main seating area. p. 156
(8) All walls should be dispersive or absorbent, or both, as required to reduce echoes or near-echoes in relation to the variety of speaking positions. In any case, wall absorbents will probably be required to control reverberation. p. 150
(9) Check carefully for the possibility of echoes from re-entrant angles. p. 157
(10) Provide sound-absorbing seating and quiet floor finishes.
(11) Walls and ceiling of galleries should be absorbent to reduce noise.
(12) If electronically amplified speech is necessary, great care must be taken in the positioning of microphones and loudspeakers. In large halls specialist advice at an early stage is essential. p. 165
(13) Reverberation time should be in the region of 1 s, depending on the size of the hall, and taking into account partial attendances. p. 163

Law courts

(1) The customary seating arrangement of law courts takes into account the need to bring the participants as near to each other as possible, for visual and acoustic reasons. This characteristic should therefore be retained.
(2) Any difficulty of hearing in law courts is usually the result of excessive reverberation, due to hard reflecting surfaces and a large volume per person. p. 159

(3) The ceiling should form a horizontal reflector and be as low as possible, consistent with the character of the room. p. 152
(4) A low angled reflector over the judge's or magistrates' bench is optional, but will assist in the control of the proceedings. p. 190
(5) The reflecting ceiling should be extended towards the public gallery, sufficiently to reinforce sound in this area. p. 152
(6) All walls should be absorbent or dispersive to reduce delayed cross-reflections. Absorbents required to control reverberation should therefore be distributed around the room on wall surfaces. p. 159
(7) The angles between walls and ceiling should be designed to prevent diagonal echoes. p. 157
(8) The walls and ceiling of the public gallery should be sound absorbing and quiet floor finishes should be used.
(9) Seating should be upholstered, preferably fabric covered, to assist in the maintenance of a low period of reverberation when the number of people present is small.
(10) The reverberation time should be not more than 1 s at a minimum room occupancy. p. 163

Lecture theatres

(1) An economic arrangement of seating and gangways is essential p. 143
(2) In larger rooms, seating rows splayed around the lecturer's dais will minimise the distance to back row. The equal intelligibility contour can be used to find the optimum proportion for the seating area, subject to satisfactory sight lines for the projector screen. p. 143
(3) Tiered seating (minimum 20° rake) is essential for good acoustics but the rake may have to be increased for good sight lines to the bottom of the blackboard, projector screen or top surface of a demonstration bench. p. 147
(4) An angled reflector and a horizontal reflecting ceiling provides good sound reinforcement from lecturer to students and vice versa, but should be kept as low as reasonably possible. p. 148
(5) The walls behind and immediately around the lecturer should act as reflectors when he turns away from his audience. p. 155
(6) Reduce cross-reflections at the dais by splays or dispersive treatment.
(7) Flank walls should be dispersive or non-parallel, and absorbent if required for the control of reverberation.
(8) The rear wall should be absorbent and if curved, concave to the audience, also dispersive. p. 159
(9) Seating or benches should be padded and, if bench fronts are employed, these should be perforated plywood with absorbent infilling. Seating will then reduce the extension of reverberation at small attendances.
(10) Prevent echoes from the angle between rear wall and ceiling. p. 157
(11) Reverberation time should be in the region of 3/4 s. p. 163

Open-air concert platforms

(1) Open-air conditions are not really appropriate for orchestral music due to the lack of reverberation.
(2) It is therefore necessary to compensate for the lack of 'body' by ensuring strong direct sound for all members of the audience, good primary reflections and a degree of localised reverberation within the orchestral 'shell'. p. 175
(3) The seating area should be tiered or as steeply raked as possible. p. 147
(4) The platform should also be tiered and of wood staging, to act as a resonator. p. 175
(5) The platform cover should be designed generally to project sound towards the seating area but with its internal surface 'broken' to provide a degree of dispersion.
(6) If the front of the platform is low, a clear paved space (or water) immediately in front will act as a reflector. Beyond this, the arrangement of seats and gangways should be economic.
(7) The shape of the seating area will be related to the plan shape of the orchestral 'shell', since the latter should project direct and reflected sound over the full width of the seating area.
(8) Protect the audience from wind and external noise by screen walls or planting, and in any case choose a quiet site. p. 89

Open-air theatres

(1) Since reflectors can be used only to a limited extent, if at all, dependence on direct sound is greater than in the covered theatre.
(2) The equal intelligibility contour should be used to determine the optimum shape for the seating area, in relation to the type of stage and the movement of actors. p. 143
(3) Seating should be as steeply banked as possible and the stage should be raised. p. 147
(4) Strict economy should be employed in the arrangement of seats and gangways to minimise the distance to the rear seats.
(5) Ideally, no seat should be more than about 20 m from the stage, on the central axis. p. 143
(6) Walls around the acting area, if practicable, will act as useful reflectors when actors turn away from the audience, even if the walls are only head height.
(7) If the front of the platform is low, a clear paved space (or water) immediately in front will act as a reflector.
(8) Protect the audience from wind as much as possible. Wind at 25 miles/h reduces by half the distance at which speech can be understood.
(9) Protect the theatre from external noise by planting or screen walls, and in any case choose a quiet site. p. 89

Opera houses

(1) The requirements of the theatre and the concert hall are combined in the design of an opera house and reference under these headings should be made.

(2) The Italian tradition of multi-gallery, horse-shoe shaped opera houses still influences the design of the modern opera house. It may be said to give it its individual character and provide an appropriate 'sense of the occasion'. The theatre fan-shape is, however, quite suitable acoustically and makes good sight lines easier to achieve.

(3) The traditional interior, in effect, provides a high degree of sound absorption on all wall surfaces, which makes an adequate degree of reverberation for Wagnerian opera difficult to attain. A reduction in the number of galleries, and their extent along flank walls, is advisable to meet this requirement and provide sight lines to modern standards.

(4) Stalls seating should be more steeply raked than has been the custom, even though this results in a reduction in the number of galleries. p. 147

(5) Provided sight lines, and therefore direct sound paths, are good, hearing is likely to be satisfactory having in mind the somewhat greater strength and clarity of the singing voice, compared with the conversational speech of the modern theatre.

(6) Galleries should be kept shallow to avoid sound shadows and highly absorbent volumes. p. 147

(7) As much assistance as possible should be given to the singers by the provision of an overhead reflector. This will be approximately over the orchestral pit and may be slightly convex if required to reflect sound towar upper galleries.

(8) The ceiling is likely to be too high to be effective as a reflector and should be dispersive to reduce near-echoes and contribute to reverberation. p. 159

(9) Even shallow domes at high level may cause serious concentrations of delayed reflections. p. 158

(10) Rear walls, if exposed to direct sound, should be made absorbent. Echoes can also be produced by reflections from re-entrant angle and balcony aprons. p. 157

(11) Flank walls, where not occupied by galleries or boxes, should be dispersive.

(12) The provision of an apron stage has two advantages. It makes it possible for singers to approach nearer the audience, making an overhead reflector more effective, and it provides in itself a reflector when singers are 'up-stage'.

(13) To provide an apron stage, the orchestral pit may be partly under the stage. Both stage and pit floor should act as resonators. p. 175

(14) Seating should be as absorbent as possible to reduce reverberation for small audiences and at rehearsals.

(15) Reverberation time should be in the region of 1.5 s. p. 163

School halls

(1) The acoustic design of school halls is made difficult by the same three factors as are mentioned under 'Community centre halls', namely

(i) the variety of uses requiring different periods of reverberation
(ii) the considerable variation in the occupancy of the room
(iii) the need to provide a level floor for dancing, games, parties, etc. p. 176

(2) Variation in occupancy may be greater than in other muli-purpose halls. Reverberation must be kept down if the room is not to be noisy when used by small groups.
(3) A completely plain horizontal ceiling can give rise to strong inter-reflections between floor and ceiling when the hall is cleared of chairs. After provision has been made for an overhead reflector near the stage, the remainder of the ceiling should be absorbent. The reflector should be kept low and set at an angle, to reduce its size. p. 153
(4) The average school hall is required to be rectangular to satisfy its various uses, although some associated with large schools have been designed on a fan shape. These reduce the distance of rear seats from the stage. Even the rectangular halls are now wider than previously, with a distinct advantage acoustically and from the point of view of sight lines from a level floor. p. 147
(5) If the whole floor is not required to be level, a raised or tiered floor at the rear will improve acoustics and sight lines. p. 147
(6) A shallow gallery will increase the proportion of seats having good sight lines and satisfactory acoustics. When occupied by school children the gallery can be a source of noise. Walls, floor and ceiling should be sound-absorbing.
(7) Flank walls should be dispersive, with areas of sound absorbing material as required to control reverberation. Piers, window reveals and curtains often meet this requirement adequately.
(8) The wall opposite the stage should be absorbent from dado level to ceiling. p. 156
(9) For dramatic performances, an apron stage has the advantage of bringing young voices out of the absorbent area of scenery and curtains, and underneath the stage reflector.
(10) To provide resonance for musical performances, the stage, including the apron, should be of wood construction. p. 175
(11) The intelligibility of speech being most critical, reverberation should not be more than 1 s for full audience conditions and, after calculation for small occupancies, may have to be less. A low period of reverberation will be suitable for films, games and dances and only the less frequent musical performances will suffer. p. 163

Theatres

(1) The shape of the seating area should be related to the type of stage adopted, by the use of equal intelligibility contours. p. 143

(2) Seating and gangway arrangements should be as economical as possible to reduce the distance from the stage to rear rows of seats.

(3) Subject to sight lines, a wide rather than deep auditorium brings the audience nearer the stage. For the proscenium type of stage the fan shape reduces the distance to rear seats to the minimum, for a given angle of sight lines. For the 'thrust stage' this distance can be further reduced by the provision of seats at the sides, provided they lie within the equal intelligibility contours. p. 145

(4) Galleries also reduce the distance to rear seats but they should not be so deep as to give rise to sound shadows for the seats below them. p. 157

(5) For modern acting technique the furthest seats should not, on the central axis, be more than 25 m from the centre of the acting area. On other axes this distance should be less, as indicated by the use of the equal intelligibility contour. It should also be borne in mind that expressions on actors' faces are hardly seen beyond a distance of about 20 m. p. 143

(6) An apron stage is desirable, not only to encourage acting beneath the proscenium reflector, but also because it acts as a reflector for 'up-stage' positions.

(7) Stalls seating should be tiered at an angle of approximately 20° or to provide clear sight lines to the front edge of the stage, whichever is the greater angle.

(8) Galleries should also provide clear sight lines to the feet of the actors in the extreme 'down-stage' position.

(9) Overhead reflectors, including the ceiling, should be designed to provide progressively increasing reinforcement of sound towards the rear of the auditorium. Reflectors should be as low as is practicable. p. 153

(10) For 'proscenium frame' productions, it should be remembered, in designing the first reflector, that the 'false proscenium opening'* is seldom more than 6 m high and must not be allowed to prevent sound reaching the reflector. p. 151

(11) Surfaces not used as reflectors should be dispersive.

(12) Rear walls above head level should be absorbent and if curved (concave to the auditorium) should also be dispersive. p. 159

(13) Balcony aprons and any other minor surfaces facing the stage should be absorbent. p. 157

(14) Prevent echoes from re-entrant angles at the rear of the auditorium. Such angles can occur in plan as well as section. p. 157

(15) Seating should be as absorbent as possible.

(16) Reverberation should be 1 to 1.25 s, depending on the size of the theatre. p. 163

Theatre-in-the-round

(1) Such theatres are generally smaller than the traditional theatre, being less commercially successful or designed specifically for limited audiences.

*The 'false proscenium' is an adjustable border behind the structural proscenium opening.

(2) In any case, the number of rows of seats around the stage is strictly limited by the fact that an actor always has his back turned to part of his audience. The equal intelligibility contour can be used to arrive at the possible number of rows of seats, for a given criterion of audibility. p. 146
(3) Economy in the arrangement of seats and gangways is just as important as in the traditional theatre.
(4) The shape of the auditorium will be largely determined by the shape of the stage and the positions of entrances for actors and public. An approximately square shape is logical acoustically, equalising the distance from stage to rear seats. If a circular shape is adopted the need to prevent concentrated echoes becomes greater. p. 150
(5) In any case, it will be necessary to treat all walls with absorbent material and possibly make them dispersive as well, if they are curved. p. 159
(6) A low horizontal or saucer-shaped reflector over the stage will, to some extent, reflect an actor's voice 'over his shoulder'. The remainder of the ceiling should be dispersive.
(7) The stage is generally required to be about level with the floor for the front row of seats. This means that the seating should be more steeply tiered than for the stalls of the traditional theatre. p. 147
(8) Seating should be as absorbent as possible.
(9) Reverberation time should be about 3/4 s. p. 163

Town halls

(1) The acoustic design of town halls is made difficult by the same three factors as are mentioned under the heading 'Community centre halls', namely

- (i) the variety of uses requiring different periods of reverberation
- (ii) the considerable variation in the potential occupancy of the room
- (iii) the need to provide a level floor for dances. p. 176

(2) Town halls vary considerably in the uses to which they are put and these must be carefully studied, and their relative importance weighed, before any decision is made regarding shape and acoustic design.
(3) Most town halls in this country have been designed on a traditional rectangular pattern quite unsuitable for good acoustics and sight lines when used for drama, meetings and lectures, and not too satisfactory for music.
(4) The disadvantages resulting from the need for a level floor are greater in a town hall than in a community centre hall because of its, normally, greater size. In view of the fact that more money is usually available for a civic centre, the architect should strongly advise his clients to provide two halls, one for functions requiring a level floor and the other as a well-designed auditorium having a tiered or raked floor.
(5) If the above is not possible, then the notes under the heading 'Community centre halls' apply equally to town halls. p. 182

Village halls

See 'Community centre halls', on p. 82.

Reverberation Calculation: Example

Figure 4.30 shows the plan and section of a conference hall to seat 380 people. This will be used to provide an example of a reverberation time calculation. It will also serve as an example of the application of the design principles outlined on p. 183 under the heading 'Conference halls'.

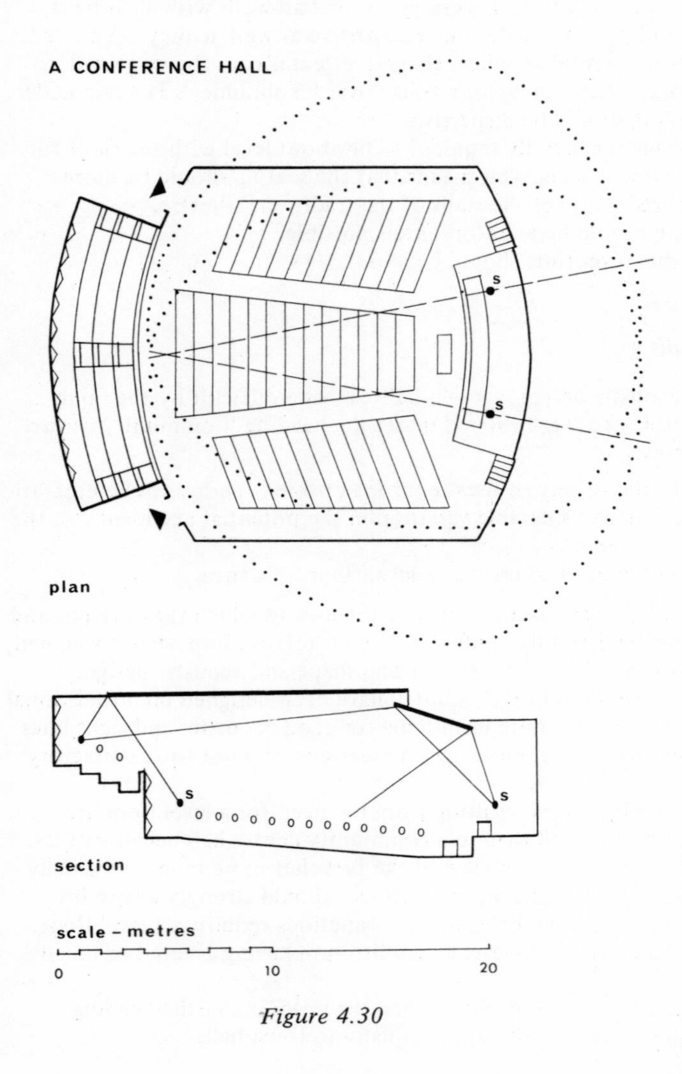

Figure 4.30

Table 4.1 illustrates the setting out of the calculation, which is largely self-explanatory. Before commencing the calculation, however, the *required* total room absorption was obtained by inverting the Sabine formula and inserting the known values. These are as follows (see figure 4.22)

Room volume as designed: $2240\,\mathrm{m}^3$

Optimum reverberation times: 1.4 s at 125 Hz
: 0.9 s at 500 and 2000 Hz

Thus

$$A = \frac{V}{\mathrm{RT}} \times 0.1608$$

$$A = \frac{2240}{1.4} \times 0.1608$$

$$= 257\,\mathrm{m}^2 \text{ sabins at 125 Hz}$$

and

$$A = \frac{2240}{0.9} \times 0.1608$$

$$= 400\,\mathrm{m}^2 \text{ sabins at 500 and 2000 Hz}$$

As will be seen, the materials chosen for the various surfaces in the hall, together with the estimated absorption of people and seats, provide total absorption figures fairly close to those required. It will also be seen that by providing for well-upholstered, fabric-covered seats the variation from optimum reverberation times due to changes in room occupancy is small.

Since, as will be explained below, such a calculation must be approximate and average for the room as a whole, the calculated reverberation times are given to only one place of decimals. To express reverberation time to more than one place of decimals would suggest an accuracy which does not exist.

Reverberation Calculation: Limits of Accuracy

The simple Sabine formula employed above is based on certain assumptions which do not always apply in practice and which give rise to some inaccuracy. The most important assumption is that there will be a diffuse and random scattering of sound within the enclosure, which is seldom the case. This and other factors which give rise to inaccuracy are examined below but it will be seen that some inaccuracies can be reduced in design when their causes are understood.

(1) The Sabine formula proves to be slightly inaccurate for cases where the total absorption of a room is high in relation to volume. Such is the case in some kinds of recording studios and acoustic laboratories. In the extreme case this can be understood by considering the hypothetical case of an enclosure where all

TABLE 4.1 A Conference Hall: Reverberation Calculation

Item and description	*Volume, area or number*	*Coeff. 125 Hz*	*Metric sabins*	*Coeff. 500 Hz*	*Metric sabins*	*Coeff. 2000 Hz*	*Metric sabins*
(1) Air (absorption negligible below 1000 Hz)	2240	–	–	–	–	0.0066	15
(2) Main floor, carpet on felt, on screed, reduction for shading	231	0.10	23	0.15	35	0.15	35
(3) Platform, carpet on wood staging	57	0.15	9	0.20	11	0.20	11
(4) Platform apron, wood panelling	14	0.25	4	0.10	1	0.05	1
(5) Gallery floor, including risers, as (2) above	70	0.10	7	0.15	11	0.15	11
(6) Platform wall, curtains over projection screen and panelling	105	0.10	11	0.20	21	0.30	32
(7) Flank walls, spaced wood strips on battens, absorbent in air space, boldly modelled	195	0.25	49	0.35	68	0.35	68
(8) Gallery apron wall, including doors, as (7) but additional absorption and boldly modelled	57	0.25	14	0.65	37	0.65	37
(9) Wall above doors, as (8) above	30	0.25	8	0.65	20	0.65	20
(10) Flank walls of gallery, plaster on solid backing	32	0.02	1	0.02	1	0.04	1
(11) Rear wall of gallery, as (7) above	45	0.25	11	0.35	16	0.35	16
(12) Main ceiling, suspended, plaster on metal lath	252	0.02	5	0.03	8	0.04	10
(13) Gallery ceiling, perforated absorbent plaster tiles	35	0.35	12	0.55	19	0.55	19
(14) Angled reflector, veneered blockboard on wood framing	52	0.15	8	0.10	5	0.10	5

TABLE 4.1 (Continued)

Item and description	*Volume, area or number*	*Coeff. 125 Hz*	*Metric sabins*	*Coeff. 500 Hz*	*Metric sabins*	*Coeff. 2000 Hz*	*Metric sabins*
(15) Ceiling over platform, plaster on solid backing	30	0.02	1	0.02	1	0.04	1
(16) Ventilation grilles, 50% voids	20	0.15	3	0.35	7	0.40	8
(17) Seating, fabric-covered upholstery	380	0.19	72	0.28	106	0.28	106
Total permanent absorption			238		349		396
(18) Audience and speakers, (coeff. additional to seating)	380	0.02	8	0.18	68	0.23	87
(19) Audience and speakers, half occupancy	190	0.02	4	0.18	34	0.23	44
Total absorption, full occupancy			246		417		483
Total absorption, half occupancy			242		383		440
Reverberation time, hall empty (s)		1.5		1.0		0.9	
Reverberation time, hall half occupied (s)		1.5		0.9		0.8	
Reverberation time, hall fully occupied (s)		1.5		0.9		0.7	
Optimum reverberation time (volume 2240 m^3) (s)		1.4		0.9		0.9	

surfaces are totally absorbent—where reverberation could not occur. Yet, by application of the Sabine formula a small but positive degree of reverberation would be indicated. In the case of highly absorbent rooms having a reverberation time of much less than 1 s the Eyring formula can be used.

(2) In rooms in which angled reflectors are used to project sound as much as possible towards the audience (a very absorbent 'surface') it is found that actual reverberation is shorter than calculated. Again, this can be confirmed by considering an even more hypothetical case of an enclosure of paraboloid shape in which the source of sound is at its focus and the paraboloid is closed by a totally absorbent surface. All direct and reflected sound would be projected at the absorbent surface and reverberation could not occur. Yet, by application of the Sabine or any other formula a positive period of reverberation would result. For this and other reasons which have been mentioned earlier, the over-employment of directional reflectors is inadvisable, especially for music.

(3) Reference has been made earlier to standing wave phenomena. In rectangular rooms with opposing reflective surfaces 'modes of resonance' occur at frequencies related in their wavelengths to the dimensions of the room. These have the effect of accentuating certain frequencies and delaying their decay. In the case of music, distortion can be noticeable, especially in smaller rooms or within rectangular recesses. The avoidance of parallel, plain and reflective surfaces is always advisable. The calculation of reverberation time will not reveal the preferential extension of certain frequencies.

(4) The Sabine and other formulae assume the diffuse reverberation resulting from the uniform and random scattering of sound within the enclosure, and where the dimensions of the enclosure do not vary to a large extent. In order to make a reasonably close estimate of reverberation time, and for other reasons which have been mentioned, it is recommended that room proportions should not be elongated and that, particularly for music, as many surfaces as possible should be made dispersive, that is, boldly modelled. The avoidance, if possible, of large changes in volume (as illustrated in figure 4.26) will also contribute to a diffuse and uniform reverberation. The calculated reverberation time can only provide an average period for the auditorium as a whole.

(5) The absorption coefficient of a material varies according to the angle of incidence of the sound and the surface area of the tested sample. Published coefficients of absorption are based on a fairly standardised method of testing under conditions of random sound incidence and for panels of about 6 m^2. In an auditorium these conditions may not apply and the absorption effectiveness of the materials employed can be different from those assumed in the calculation. By an understanding of the conditions under which materials are tested some allowance can be made when they are employed in a markedly different manner.

(6) A practical problem encountered in calculating reverberation time is due to the difficulty, in all but the simplest interiors, of measuring all the intricate surfaces which may be exposed to sound. Actual surface areas may be greatly increased, and resonant pockets fortuitously provided, by complications in design which are difficult to measure. Furnishings present another problem since absorption factors are only available for chairs and theatre seats. It is essential therefore to measure room surfaces as accurately as possible, to measure the surface area of a modelled surface, not its elevational area, to make due

allowance for such incidental items as ventilation grilles and to keep in mind that reverberation is likely to be slightly shorter than calculated.

(7) The absorption factors published for members of an audience normally assume a closely seated arrangement. If there are gaps between members of the audience absorption per person will be increased owing to a lack of screening of one person by another (see 'Sound absorption factors' on p. 202). Thus, whereas calculation of reverberation time for a full audience or for an empty auditorium is not invalidated by this factor, the calculated time for a partial audience is less reliable owing to their unpredictable disposition. In fact, RT is likely to be shorter than calculated, especially at high frequencies.

Although it will be seen from the above considerations that calculation can only approximate actual reverberation time when the room is tested, it will also have been seen that errors can be minimised. In the absence of calculations errors in design can be far more serious than any errors in the method of calculation.

It is, however, with a view to correcting these errors that allowance is sometimes made for 'tuning-in' an auditorium after it is built by the provision of areas where the surface material can be modified. For the same reason a technique has been developed for the making of scale models by which acoustic design can be checked before actual building takes place.

Acoustic Models

The technique of making scale models of auditoria into which speech or music can be introduced and results checked, is a comparatively recent development. It is an expensive operation which can probably only be justified in the case of expensive buildings such as theatres, opera houses, concert halls and conference centres. In any event, the making of a model does not obviate careful and knowledgeable design. If time and money are not to be wasted, the designer must get as near as possible to a good solution to his problem before the model is built.

Since the behaviour of sound in an enclosure is related to the wavelengths of the frequencies involved, the wavelengths of the sound introduced into a model must be scaled down to correspond with the reduced scale of the model. Thus when music is introduced into the model all frequencies have to be raised by a factor of, say, 8 if the model is one-eighth full-size. Furthermore, materials have to be used for the interior of the model which will coincide in their patterns of absorption coefficients with those it is intended to use—but at the higher range of frequencies. This correlation will also apply to the 'model audience' and even the air within the model, which is dehydrated for the purpose.

Figure 4.31 shows in a simplified form the various stages in the technique which can be summarised as follows with reference to the numbers on the drawing

(1) Music is played in a non-reverberant interior and
(2) a microphone picks up the sound which is
(3) recorded on magnetic tape.
(4) All frequencies are raised by a factor of 8 and
(5) played into the one-eighth scale model.

(6) The sound, modified by the enclosure, is then picked up by probe microphones and

(7) re-recorded.

(8) All frequencies are then reduced to their natural level by a factor of one-eighth and

(9) the music is played to the listener in a non-reverberant room for a subjective appraisal of the result.

From such a test it should be possible to detect faults such as echoes and standing wave phenomena and compare the amount and quality of reverberation in various parts of the auditorium.

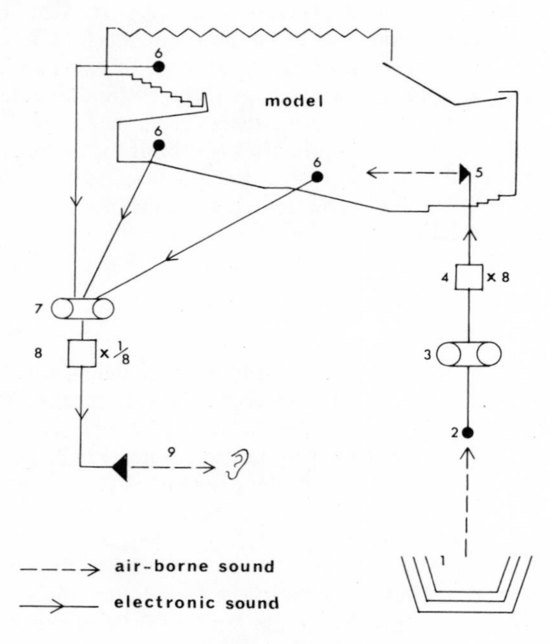

Figure 4.31

Coefficients of Sound Absorption

In the following schedule composite materials are listed alphabetically according to their surface material. Coefficients can be multiplied by the area in square metres or square feet to give absorption in square metre sabins or square foot sabins. Absorption factors for air, people and seats are listed separately at the end of the schedule.

		125	500	2000 Hz
Asbestos	13 mm sprayed on solid backing	0.10	0.40	0.80
Asbestos	25 mm ditto	0.15	0.50	0.80

		125	500	2000 Hz
Asbestos tiles	5 mm perforated asbestos cement on battens, 25 mm mineral wool in air space	0.17	0.94	0.65
Asbestos tiles	ditto but 50 mm mineral wool in air space	0.29	0.94	0.67
Asbestos tiles	3 mm perforated asbestos cement backed with 25 mm rock-wool pad on battens, 25 mm free air space	0.32	0.72	0.76
Audience area	audience area including gangways not more than 1.1 m wide	0.52	0.85	0.93
Brickwork	open texture, unpainted	0.02	0.03	0.04
Brickwork	dense texture, unpainted	0.01	0.02	0.03
Brickwork	painted two coats	0.01	0.02	0.03
Carpet	thin, on thin felt, on solid floor	0.10	0.25	0.30
Carpet	thick, on thick felt, on solid floor	0.10	0.50	0.60
Carpet	thin, on thin felt, on boards and joists or battens	0.20	0.30	0.30
Chairs	see under Seating			
Clinker blocks	unplastered	0.20	0.60	0.50
Concrete	rough finish	0.02	0.02	0.05
Concrete	smooth finish	0.01	0.02	0.02
Cork tiles	19 mm floor tiles bedded solid, sealed and polished	0.02	0.05	0.10
Curtains	in folds against wall, medium weight	0.10	0.40	0.50
Curtains	dividing spaces, medium weight	0.03	0.10	0.20
Fibreboard	13 mm wood or cane fibre, soft texture, solid mounting	0.05	0.15	0.30
Fibreboard	ditto but two coats emulsion	0.05	0.10	0.10
Fibreboard	13 mm wood or cane fibre, soft texture, on battens, 25 mm free air space	0.30	0.35	0.30
Fibreboard	ditto but two coats emulsion	0.30	0.15	0.10
Fibreboard	19 mm wood fibreboard with 75 mm grooves	0.21	0.46	0.89
Fibreboard	6.4 mm perforated soft wood fibreboard over channelled soft wood fibreboard	0.18	0.46	0.89
Fibreboard	12.5 mm soft wood fibreboard perforated through entire thickness	0.07	0.75	0.38
Fibreboard	12.5 mm soft wood fibreboard channelled 6.4 mm deep	0.14	0.36	0.46
Fibreboard	19 mm soft wood fibreboard channelled 8 mm deep	0.19	0.46	0.82
Fibreboard	15 mm mineral fibre, textured and perforated	0.54	0.40	0.59
Fibreboard	15 mm mineral fibre, striated surface, suspended, 300 mm cavity	0.61	0.43	0.60
Fibreboard	15 mm mineral fibre, fissured surface, suspended, 300 mm cavity	0.35	0.40	0.80
Fibreboard	15 mm mineral fibre, fissured surface 22 mm air space	0.09	0.86	0.90

Material	Description	125	500	2000 Hz
Fibreboard	19 mm mineral fibre, heavy texture, suspended, 400 mm cavity	0.06	0.73	0.88
Fibreboard	Ditto, solid mounting	0.04	0.73	0.88
Fibreboard	19 mm cross-channelled fibreboard on battens, 25 mm air space	0.20	0.65	0.88
Fibreboard	13 mm perforated wood fibreboard on battens, 25 mm air space	0.20	0.60	0.65
Glass	in windows, up to 4 mm	0.30	0.10	0.07
Glass	ditto but 6 mm	0.10	0.05	0.02
Glass	bedded solid	0.01	0.01	0.02
Glass fibre	25 mm resin bonded	0.10	0.55	0.75
Glass fibre	ditto but 50 mm	0.20	0.70	0.75
Granolithic	granolithic cement screed	0.01	0.02	0.02
Hardboard	panelling on battens, 25 mm air space	0.20	0.15	0.10
Hardboard	10% perforated panelling, 25 mm fibre-glass in air space	0.10	0.55	0.80
Hardboard	ditto but 50 mm fibreglass	0.20	0.75	0.70
Hardboard tiles	perforated, on 19 mm channelled soft fibreboard, on battens, 25 mm air space	0.21	0.66	0.80
Linoleum	on solid floor	0.05	0.05	0.10
Marble	on solid backing	0.01	0.01	0.02
Metal	3% perforated panelling, 25 mm fibreglass in air space	0.10	0.70	0.40
Metal	ditto but 50 mm fibreglass	0.20	0.90	0.25
Metal	10% perforated panelling, 25 mm fibreglass in air space	0.10	0.55	0.80
Metal	ditto but 50 mm fibreglass	0.20	0.75	0.70
Metal	perforated trays with 32 mm mineral wool pads, 50 mm air space	0.20	0.80	0.80
Metal	perforated stove-enamelled metal trays, 64 mm mineral wool pads, free air space	0.25	0.96	0.85
Openings grilles	ventilation grilles, 50% voids	0.15	0.35	0.40
Openings stage	proscenium opening area, average curtains or scenery	0.20	0.30	0.40
Openings windows	elevational area of opening	1.00	1.00	1.00
Plaster	lime or gypsum on solid backing	0.03	0.02	0.04
Plaster	ditto on lath, small air space	0.30	0.10	0.04
Plaster	suspended on metal lath, large air space	0.20	0.10	0.04
Plaster	13 mm acoustic plaster on solid backing	0.13	0.35	0.45
Plasterboard	on firring, small air space	0.03	0.01	0.04
Plasterboard	12% perforated, 25 mm mineral wool and 25 mm air space	0.17	0.90	0.45
Plasterboard	12% perforated, 50 mm mineral wool and 38 mm air space	0.37	0.85	0.45

		125	500	2000 Hz
Plaster tiles	19 mm perforated fibrous plaster, aluminium foil backing, free air space	0.45	0.80	0.65
Plaster tiles	16 mm perforated gypsum tiles, 25 mm fibreglass quilt, 25 mm air space	0.30	0.80	0.45
Plaster tiles	25 mm perforated asbestos fibre with skim coat plaster, 25 mm air space	–	0.75	–
Plaster tiles	25 mm acoustic plaster tiles on solid backing	0.15	0.40	0.50
Plastic cloth	perforated plastic cloth on 19 mm perforated wood fibreboard, on battens, 25 mm air space	0.30	0.70	0.75
Plastic tiles	plastic floor tiles	0.05	0.05	0.10
Plastic tiles	13 mm expanded polystyrene, unperforated, 25 mm air space	0.05	0.40	0.20
Plastic tiles	19 mm perforated expanded polystyrene, 25 mm air space	0.05	0.70	0.20
Rubber	flooring, on solid floor	0.05	0.05	0.10
Seating area	fabric covered seating with perforated under-seats, area inclusive of gangways not more than 1.1 m wide	0.44	0.77	0.82
Shading	reduction of coefficient for floor finish where floor is 'shaded' by seating	20%	40%	60%
Stage	see under Wood staging			
Steel	see under Metal			
Stone	polished	0.01	0.01	0.02
Stone	dressed	0.02	0.02	0.04
Tapestry	stretched over 38 mm air space	0.05	0.25	0.30
Terrazzo	on walls or floors	0.01	0.01	0.02
Tiles	glazed	0.01	0.01	0.02
Tiles	unglazed	0.03	0.03	0.05
Ventilation grilles	see under Openings			
Windows	see under Glass and Openings			
Water	as in swimming pool	0.01	0.01	0.02
Wood blocks	on solid floor	0.02	0.05	0.10
Wood boarding	25 mm boards on joists or battens, free air space	0.10	0.10	0.05
Wood boarding	19 mm match boarding on battens on solid wall	0.30	0.10	0.10
Wood panels	3 ply panelling on solid back	0.10	0.10	0.05
Wood panels	3 ply panelling on battens, 25 mm free air space	0.30	0.15	0.10
Wood panels	3 ply panelling on battens, 25 mm acoustic felt in air space	0.40	0.15	0.10
Wood panels	ditto but 50 mm acoustic felt	0.50	0.25	0.20

		125	500	2000 Hz
Wood panels	15% perforated 3 ply panelling on battens, 50 mm fibreglass in air space	0.20	0.60	0.35
Wood panels	ditto with 25 mm free air space in addition	0.33	0.80	0.40
Wood panels	ditto but 5% perforated	0.40	0.63	0.13
Wood staging	25 mm boarding on staging with large air space	0.15	0.10	0.10
Wood strips	38 mm wood strips spaced to give 5 mm slots, 25 mm mineral wool in air space	0.25	0.65	0.65
Wood veneer tiles	wood veneered fibreboard tiles, perforated, on battens, 25 mm air space	0.20	0.50	0.75
Wood strips	13 mm profiled and slotted cedar on battens, 38 mm mineral wool in air space	0.20	0.98	0.21
Wood strips	Ditto but without mineral wool	0.04	0.36	0.16
Woodwool	25 mm woodwool slabs, unplastered, solid backing	0.10	0.40	0.60
Woodwool	ditto but 75 mm	0.20	0.80	0.80
Woodwool	25 mm woodwool slabs, unplastered, on battens, 25 mm air space	0.10	0.60	0.60

Sound Absorption Factors

		125	500	2000 Hz
Air	average temperature and humidity:			
	per m^3	–	–	0.0066
	per ft^3	–	–	0.002
Audience	closely spaced, per person,			
	m^2 sabins	0.21	0.46	0.51
	ft^2 sabins	2.30	5.00	5.50
Audience	seated about two metres apart, per person,			
	m^2 sabins	0.21	0.53	0.97
	ft^2 sabins	2.30	5.70	10.40
Seating	wood, per chair,			
	m^2 sabins	0.01	0.02	0.02
	ft^2 sabins	0.15	0.17	0.20
Seating	metal and canvas, per seat,			
	m^2 sabins	0.07	0.15	0.18
	ft^2 sabins	0.80	1.60	1.90
Seating	leather padded, per seat			
	m^2 sabins	0.11	0.15	0.18
	ft^2 sabins	1.20	1.60	1.90
Seating	leather upholstered, with perforated under-seat, per seat,			
	m^2 sabins	0.19	0.23	0.23
	ft^2 sabins	2.00	2.50	2.50

		125	500	2000 Hz
Seating	fabric upholstered, with perforated under-seat, per seat,			
	m^2 sabins	0.19	0.28	0.28
	ft^2 sabins	2.00	3.00	3.00
Seating	heavily upholstered, fabric covered, per seat,			
	m^2 sabins	0.26	0.28	0.30
	ft^2 sabins	2.80	3.00	3.30

Note: The metric values above are for use in the metric formula given on p. 131. The imperial values are for use only in the formula given at the foot of p. 164.

The sound absorption of *areas* of audience or seating can also be measured by using the coefficients in the main part of the schedule. See under Audience area and Seating area.

Metric–Imperial Conversion

To convert metres into feet multiply by 3.281
To convert square metres into square feet multiply by 10.764
To convert cubic metres into cubic feet multiply by 35.319
To convert feet into metres multiply by 0.305
To convert square feet into square metres multiply by 0.093
To convert cubic feet into cubic metres multiply by 0.028

Absorption coefficients are not affected. Absorption factors in metric and imperial units are given on p. 202.

The Sabine formula on pp. 131 and 164 is

$$RT = \frac{\text{volume in m}^3}{\text{Absorption in m}^2\text{ sabins}} \times 0.1608$$

If measurement is imperial the formula is

$$RT = \frac{\text{volume in ft}^3}{\text{Absorption in ft}^2\text{ sabins}} \times 0.049$$

The screening formula on p. 125 is not affected

$$R = 8.2 \log_{10} \frac{44H}{\lambda} \times \tan \frac{\theta}{2}$$

where H and λ are either in metres or feet.

Figure 4.32 provides a graph for the approximate conversion of metres, square metres and cubic metres into their corresponding imperial values, and vice versa.

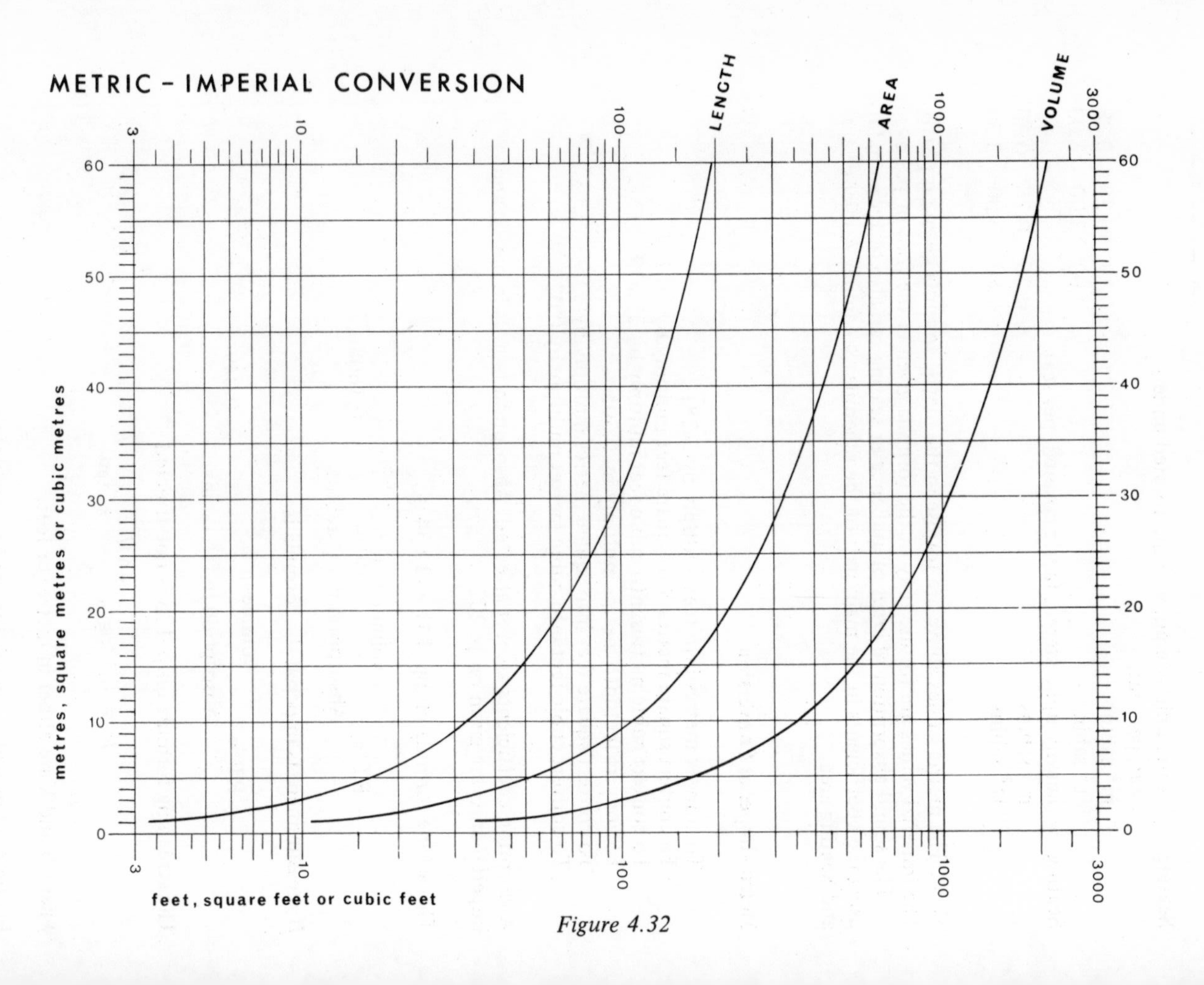
METRIC - IMPERIAL CONVERSION
LENGTH
AREA
VOLUME
metres, square metres or cubic metres
feet, square feet or cubic feet
0
10
20
30
40
50
60
3
10
100
1000
3000

Figure 4.32

Bibliography

The following is a list of publications (in date order) which will be useful to the student who wishes to delve more deeply into a particular aspect of the subject. It does not pretend to be a complete bibliography nor is it implied that any books that do not appear in it are not well worth while reading.

Influence of a Single Echo on the Audibility of Speech, E. Meyer (Department of Scientific and Industrial Research, London, 1949).

Sound-absorbing Materials, C. Zwikker and C. W. Kosten (Elsevier, Amsterdam, 1949).

Acoustical Designing in Architecture, V. O. Knudsen and C. M. Harris (Wiley, New York, 1950).

Sound Insulation and Room Acoustics, P. V. Bruel (Chapman & Hall, London, 1951).

Acoustics in Modern Building Practice, F. Ingerslev (Architectural Press, London, 1952).

Acoustics, L. L. Beranek (McGraw-Hill, New York, 1954).

Noise Measurement Techniques, National Physical Laboratory (H.M.S.O., 1955).

Acoustics for the Architect, A. Burris-Meyer and L. S. Goodfriend (Reinhold, New York, 1957).

Handbook of Noise Control, ed. C. M. Harris (McGraw-Hill, New York, 1957).

Noise in Factories, A. G. Aldersey-Williams (H.M.S.O., 1960).

Noise Reduction, L. L. Beranek (McGraw-Hill, New York, 1960).

Sound-absorbing Materials, E. J. Evans and E. N. Bazley (H.M.S.O., 1961).

Control of Noise, Conference Proceedings, National Physical Laboratory (H.M.S.O., 1962).

Music, Acoustics and Architecture, L. L. Beranek (Wiley, New York, 1962).

Acoustics, Noise and Buildings, P. H. Parkin and H. R. Humphreys (Faber, London, 1963).

Noise Measurement and Control, ed. P. Lord and F. L. Thomas (Heywood, London, 1963).

Traffic in Towns: Buchanan Report, C. Buchanan (H.M.S.O., 1963).

'The Noise Problem of Buildings near Airports', E. F. Stacy, *Ann. occup. Hyg.*, 3 (1963) pp. 94-106.

Noise: Wilson Report(H.M.S.O., 1963).

Noise in Industry, D. E. Broadbent (Department of Scientific and Industrial Research, London, 1964).

Room and Building Acoustics and Noise Abatement, W. Furrer (Butterworth, London, 1964).
Transport and Town Planning, H. J. Purkis (H.M.S.O., 1964).
Measurement and Suppression of Noise, A. J. King (Chapman & Hall, London, 1965).
Airborne Sound Insulation of Partitions, E. N. Bazley, National Physical Laboratory (H.M.S.O., 1966).
Building Physics–Acoustics, H. J. Purkis (Pergamon, Oxford, 1966).
Noïse Control in Industry, W. A. Hines (Business Publications, Homewood, Ill., 1966).
Traffic Noise, Greater London Council (H.M.S.O., 1966).
Airborne, Impact and Structure-borne Noise, U.S. Department of Housing (U.S. Government Printing Office, 1967).
Factory Buildings–the Noise Problem in Factory Design, J. E. Moore, ed. E. D. Mills (Morgan-Grampian, London, 1967).
The Law on Noise, Noise Abatement Society (London, 1969).
Architectural Acoustics, A. Lawrence (Elsevier, Amsterdam, 1970).
Assaults on Our Senses, J. Barr (Methuen, London, 1970).
Effects of Noise on Man, K. D. Kriter (Academic Press, New York, 1970).
Introduction to Acoustics, R. D. Ford (Elsevier, Amsterdam, 1970).
Noise, R. Taylor (Pelican Series, Penguin, Harmondsworth, 1970).
Noise Abatement, C. Duerden (Butterworth, London, 1970).
Sound, Noise and Vibration Control, L. F. Yerges (Reinhold, New York, 1970).
Building Acoustics, ed. T. Smith *et al.* (Oriel, Stockfield, 1971).
Sound Insulation in Buildings, H. R. Humphreys and D. J. Melluish (H.M.S.O., 1971).
Building Regulations 1972 (H.M.S.O.).
Code of Practice for Reducing Exposure of Employed Persons to Noise, Department of Employment (H.M.S.O., 1972).
Concepts in Architectural Acoustics, M. Egan (McGraw-Hill, New York, 1972).
Environmental Acoustics, L. L. Doelle (McGraw-Hill, New York, 1972).
Noise and the Worker, Department of Employment (H.M.S.O., 1972).
Prediction of Traffic Noise Levels, M. E. Delaney, National Physical Laboratory (H.M.S.O., 1972).
Noise and Man, W. Burns (John Murray, London, 1973).
Control of Pollution Act 1974 (H.M.S.O., reprinted 1975).
The Law Relating to Noise, C. S. Kerse (Oyez, London, 1975).
Noise and Noise Control, 2 vols, M. J. Crocker and A. J. Price (CRC, Cleveland, Ohio, 1975).
Sound, Man and Building, L. H. Shaudinischky (Applied Science, Barking, 1976).
Noise, Buildings and People, D. J. Croome (Pergamon, Oxford, 1977).

Publications of the Noise Advisory Council

Neighbourhood Noise, H.M.S.O., 1971
Traffic Noise Vehicle Regulations, H.M.S.O., 1972
A Guide to Noise Units, H.M.S.O., 1973

Noise in Public Places, H.M.S.O., 1974
Aircraft Engine Noise Research, H.M.S.O., 1974
Noise in the Next Ten Years, H.M.S.O., 1974

Publications of the Building Research Establishment: BRE Digests

Sound Absorbent Treatments, Digest 36, 1951
Sound Insulation of Buildings, Digests 88 and 89, 1956
Noise in the Home, Digest 7, 1961
Noise and Buildings, Digest 38, 1963
Sound Insulation of Traditional Dwellings, Digests 102 and 103, 1969
Insulation Against External Noise, Digests 129 and 130, 1971
Motorway Noise and Dwellings, Digest 135, 1971
Motorway Noise and Dwellings, Digest 153, 1973
Sound Insulation, Basic Principles, Digest 143, 1976
Prediction of Traffic Noise, Digests 185 and 186, 1976
Noise Abatement Zones, Digests 203 and 204, 1977

Publications of the Building Research Establishment: Current papers

Instruments for Noise Measurement, 1964
Subjective Response to Traffic Noise, 1968
Insulation Against Aircraft Noise, 1968
Traffic Noise Index, 1968
Buildings in Noisy Areas, Interaction of Acoustic and Thermal Design, 1969
Acceptable Criteria for Noise in Large Offices, 1970
Assisted Resonance in the Royal Festival Hall, 1971
Field Performance of a Noise Barrier, 1971
Designing Against Noise from Road Traffic, 1971
Simulated Sonic Boom Tests, 1971
Designing Offices Against Traffic Noise, 1973
A Survey of the Sound Insulation Between Dwellings in Modern Building Construction, 1974
Field Measurements of the Sound Insulation of Cavity Party Walls of Aerated Concrete Blockwork, 1975

Publications of the Department of the Environment

Sound Insulation in Buildings, 1971
New Housing and Road Traffic Noise, Design Bulletin 26, 1972
Planning and Noise, 1973
Calculation of Road Traffic Noise, 1975

Index